Quantities, Units and Symbols in Physical Chemistry

INTERNATIONAL UNION OF PURE AND APPLIED CHEMISTRY
PHYSICAL CHEMISTRY DIVISION
COMMISSION ON PHYSICOCHEMICAL SYMBOLS,
TERMINOLOGY AND UNITS

IUPAC

INTERNATIONAL UNION OF PURE AND APPLIED CHEMISTRY

Quantities, Units and Symbols in Physical Chemistry

Prepared
for publication by

IAN MILLS
(Chairman), *Reading*

TOMISLAV CVITAŠ
Zagreb

KLAUS HOMANN
Darmstadt

NIKOLA KALLAY
Zagreb

KOZO KUCHITSU
Tokyo

BLACKWELL SCIENTIFIC PUBLICATIONS
OXFORD LONDON EDINBURGH
BOSTON PALO ALTO MELBOURNE

Editorial offices:
Osney Mead, Oxford, OX2 0EL
(*Orders*: Tel. 0865 240201)
8 John Street, London WC1N 2ES
23 Ainslie Place, Edinburgh EH3 6AJ
3 Cambridge Center, Suite 208, Cambridge, Massachusetts 02142, USA
667 Lytton Avenue, Palo Alto California 94301, USA
107 Barry Street, Carlton Victoria 3053, Australia

First published 1988
Reprinted 1988

Printed in Great Britain at the Alden Press, Oxford and bound at Green Street Bindery, Oxford

DISTRIBUTORS

USA
Blackwell Scientific Publications Inc
P O Box 50009, Palo Alto
California 94303
(*Orders*: Tel. (415) 965-4081)

Canada
Oxford University Press
70 Wynford Drive
Don Mills
Ontario M3C 1J9
(*Orders*: Tel. (416) 441-2941)

Australia
Blackwell Scientific Publications (Australia) Pty Ltd
107 Barry Street
Carlton, Victoria 3053
(*Orders*: Tel. (03) 347 0300)

British library
Cataloguing in Publication Data

Quantities, units and symbols in physical chemistry.
1. Chemistry, Physical and theoretical—Terminology 2. Chemistry, Physical and theoretical—Notation
I. Mills, Ian II. International Union of Pure and Applied Chemistry. *Commission on Physicochemical Symbols, Terminology and Units*
541.3′014 QD451.5

ISBN 0-632-01773-2

Library of Congress
Cataloging-in-Publication Data

Quantities, units, and symbols in physical chemistry.
At head of title: International Union of Pure and Applied Chemistry, Division of Physical Chemistry, Commission on Physicochemical Symbols, Terminology, and Units.
Rev. ed. of: Manual of symbols and terminology for physicochemical quantities and units. 1979
Bibliography: p.
Includes index.
1. Chemistry, Physical and theoretical—Notation. 2. Chemistry, Physical and theoretical—Terminology. I. Mills, Ian (Ian M.) II. International Union of Pure and Applied Chemistry. Commission on Physicochemical Symbols, Terminology and Units. III. Manual of symbols and terminology for physicochemical quantities and units.
QD451.5.Q36 1987 541.3′014 87-11678

ISBN 0-632-01773-2

Contents

Preface

The objective of this manual is to improve the international exchange of scientific information. The recommendations made to achieve this end come under three general headings. The first is the use of quantity calculus for handling physical quantities, and the general rules for the symbolism of quantities and units, described in chapter 1. The second is the use of internationally agreed symbols for the most frequently used quantities, described in chapter 2. The third is the use of SI units wherever possible for the expression of the values of physical quantities; the SI units are described in chapter 3.

Later chapters are concerned with recommended mathematical notation (chapter 4), the present best estimates of physical constants (chapters 5 and 6), and conversion factors between SI and non-SI units with examples of their use (chapter 7). References (on p. 121) are indicated in the text by numbers (and letters) in square brackets.

We would welcome comments, criticism, and suggestions for further additions to this book. Offers to assist in the translation and dissemination in other languages should be made in the first instance either to IUPAC or to the Chairman of the Commission.

We wish to thank the following colleagues, who have contributed significantly to this edition through correspondence and discussion:

R.A. Alberty (Cambridge, Mass); P.R. Bunker (Ottawa); G.W. Castellan (Maryland); A. Covington (Newcastle upon Tyne); H.B.F. Dixon (Cambridge); D.H. Everett (Bristol); M.B. Ewing (London); R.D. Freeman (Oklahoma); D. Garvin (NBS, Washington); G. Gritzner (Linz, Austria); K.J. Laidler (Ottawa); I. Levine (Brooklyn, NY); D.R. Lide (NBS, Washington); J.W. Lorimer (London, Ontario); R.L. Martin (Melbourne); M.L. McGlashan (London); J. Michl (Austin, Texas); R. Parsons (Southampton); A.D. Pethybridge (Reading); M. Quack (Zurich); J.C. Rigg (Wageningen, the Netherlands); J. Rouquerol (Marseilles); G. Schneider (Bochum, W. Germany); N. Sheppard (East Anglia); K.S.W. Sing (London); G. Somsen (Amsterdam); H. Suga (Osaka); D.H. Whiffen (Stogursey, Somerset, UK).

Commission on Physicochemical Symbols, Terminology and Units

Ian Mills (Chairman)
Tomislav Cvitaš
Klaus Homann
Nikola Kallay
Kozo Kuchitsu

Historical introduction

The first edition of the 'Manual of Symbols and Terminology for Physicochemical Quantities and Units' [1.a], of which this is the direct successor, was prepared for publication on behalf of the Physical Chemistry Division of IUPAC by M.L. McGlashan in 1969, when he was Chairman of the Commission on Physicochemical Symbols, Terminology and Units. He made a substantial contribution towards the objective which he described in the preface to that first edition as being 'to secure clarity and precision, and wider agreement in the use of symbols, by chemists in different countries, among physicists, chemists and engineers, and by editors of scientific journals'. The second edition of the manual prepared for publication by M.A. Paul in 1973 [1.b], and the third edition prepared by D.H. Whiffen in 1979 [1.c], were revisions to take account of various developments in the Système International d'Unités (SI), and other developments in nomenclature.

The present volume is a substantially revised and extended version of the earlier editions, with a slightly simplified title. The decision to embark on this project was taken at the IUPAC General Assembly at Leuven in 1981, when D.R. Lide was Chairman of the Commission. The working party was established at the 1983 meeting at Lyngby, when K. Kuchitsu was Chairman, and the project has received strong support throughout from all present and past members of Commission I.1 and other Physical Chemistry Commissions, particularly D.R. Lide, D.H. Whiffen and N. Sheppard.

The extensions include some of the material previously published in appendices [1.d–k], all the newer resolutions and recommendations on units by the Conférence Générale des Poids et Mesures (CGPM) [2], and the recommendations of the International Union of Pure and Applied Physics (IUPAP) of 1978 [3] and of Technical Committee 12 of the International Organization for Standardization (ISO/TC 12) [4–6]. The tables of physical quantities (chapter 2) have been extended to include defining equations and SI units for each quantity. The style of the manual has also been slightly changed from being a book of rules towards being a manual of advice and assistance for the day-to-day use of practising scientists. Examples of this are the inclusion of extensive footnotes and explanatory text inserts in chapter 2, and the introduction to quantity calculus and the tables of conversion factors between SI and non-SI units and equations in chapter 7.

Department of Chemistry
University of Reading
July 1987

Ian Mills
Chairman
Commission on Physicochemical Symbols,
Terminology and Units

The membership of the Commission during the period 1963 to 1986, during which the successive editions of this manual were prepared, was as follows:

Titular members
Chairman: 1963–1967 G. Waddington (USA); 1967–1971 M.L. McGlashan (UK); 1971–1973 M.A. Paul (USA); 1973–1977 D.H. Whiffen (UK); 1977–1981 D.R. Lide Jr (USA); 1981–1985 K. Kuchitsu (Japan); 1985– I.M. Mills (UK).

Secretary: 1963–1967 H. Brusset (France); 1967–1971 M.A. Paul (USA); 1971–1975 M. Fayard (France); 1975–1979 K.G. Weil (Germany); 1979–1983 I. Ansara (France); 1983–1985 N. Kallay (Yugoslavia); 1985–1987 K.H. Homann (Germany); 1987– T. Cvitaš (Yugoslavia).

Members: 1975–1983 I. Ansara (France); 1965–1969 K.V. Astachov (USSR); 1963–1971 R.G. Bates (USA); 1963–1967 H. Brusset (France); 1985– T. Cvitaš (Yugoslavia); 1963 F. Daniels (USA); 1981– E.T. Denisov (USSR); 1967–1975 M. Fayard (France); 1963–1965 J.I. Gerassimov (USSR); 1979–1987 K.H. Homann (Germany); 1963–1971 W. Jaenicke (Germany); 1967–1971 F. Jellinek (Netherlands); 1977–1985 N. Kallay (Yugoslavia); 1973–1981 V. Kello (Czechoslovakia); 1985–1987 W.H. Kirchhoff (USA); 1971–1980 J. Koefoed (Denmark); 1979–1987 K. Kuchitsu (Japan); 1971–1981 D.R. Lide Jr (USA); 1963–1971 M.L. McGlashan (UK); 1983– I.M. Mills (UK); 1963–1967 M. Milone (Italy); 1967–1973 M.A. Paul (USA); 1963–1967 K.J. Pedersen (Denmark); 1967–1975 A. Perez-Masia (Spain); 1971–1979 A. Schuyff (Netherlands); 1967–1970 L.G. Sillén (Sweden); 1963–1967 G. Waddington (USA); 1981–1985 D.D. Wagman (USA); 1971–1979 K.G. Weil (Germany); 1971–1977 D.H. Whiffen (UK); 1963–1967 E.H. Wiebenga (Netherlands).

Associate members
1983– R.A. Alberty (USA); 1983–1987 I. Ansara (France); 1979– E.R. Cohen (USA); 1979–1981 E.T. Denisov (USSR); 1971–1973 W. Jaenicke (Germany); 1985– N. Kallay (Yugoslavia); 1981–1983 D.R. Lide Jr (USA); 1971–1979 M.L. McGlashan (UK); 1973–1981 M.A. Paul (USA); 1975–1983 A. Perez-Masia (Spain); 1979–1987 A. Schuyff (Netherlands); 1963–1971 S. Seki (Japan); 1969–1977 J. Terrien (France); 1975–1979 L. Villena (Spain); 1967–1969 G. Waddington (USA); 1979–1983 K.G. Weil (Germany); 1977–1985 D.H. Whiffen (UK).

1
Physical quantities and units

1.1 PHYSICAL QUANTITIES AND QUANTITY CALCULUS

The value of a *physical quantity* is equal to the product of a *numerical value* and a *unit*:

$$\text{physical quantity} = \text{numerical value} \times \text{unit}$$

Neither the physical quantity, nor the symbol used to denote it, should imply a particular choice of unit.

Physical quantities, numerical values, and units, may all be manipulated by the ordinary rules of algebra. Thus we may write, for example, for the wavelength λ of one of the yellow sodium lines:

$$\lambda = 5.896 \times 10^{-7}\ \text{m} = 589.6\ \text{nm} \quad (1)$$

where m is the symbol for the unit of length called the metre (see chapter 3), nm is the symbol for the nanometre, and the units m and nm are related by

$$\text{nm} = 10^{-9}\ \text{m} \quad (2)$$

The equivalence of the two expressions for λ in equation (1) follows at once when we treat the units by the rules of algebra and recognize the identity of nm and 10^{-9} m in equation (2). The wavelength may equally well be expressed in the form

$$\lambda/\text{m} = 5.896 \times 10^{-7} \quad (3)$$

or

$$\lambda/\text{nm} = 589.6 \quad (4)$$

In tabulating the numerical values of physical quantities, or labelling the axes of graphs, it is particularly convenient to use the quotient of a physical quantity and a unit in such a form that the values to be tabulated are pure numbers, as in equations (3) and (4).

Examples

T/K	$10^3\text{K}/T$	p/MPa	$\ln(p/\text{MPa})$
216.55	4.6179	0.5180	−0.6578
273.15	3.6610	3.4853	1.2486
304.19	3.2874	7.3815	1.9990

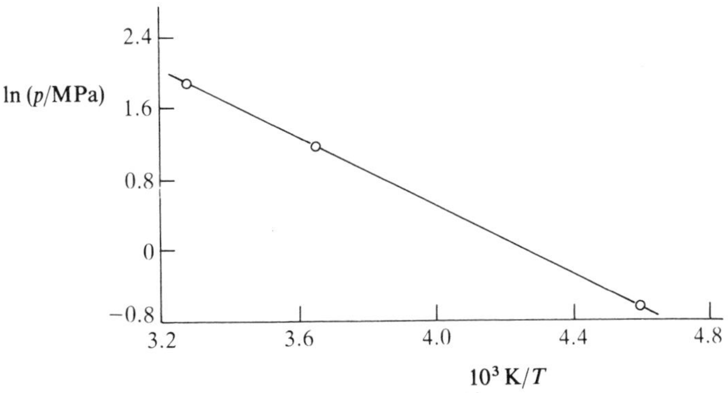

Algebraically equivalent forms may be used in place of $10^3\text{K}/T$, such as kK/T or $10^3(T/\text{K})^{-1}$.

The method described here for handling physical quantities and their units is known as *quantity calculus*. It is recommended for use throughout science and technology. The use of quantity calculus does not imply any particular choice of units; indeed one of the advantages of quantity calculus is that it makes changes between units particularly easy to follow. Further examples of the use of quantity calculus are given in chapter 7, which is concerned with the problems of transforming from one set of units to another.

1.2 BASE PHYSICAL QUANTITIES AND DERIVED PHYSICAL QUANTITIES

By convention physical quantities are organized in a dimensional system built upon seven *base quantities*, each of which is regarded as having its own dimension. These base quantities and the symbols used to denote them are as follows:

Physical quantity	*Symbol for quantity*
length	l
mass	m
time	t
electric current	I
thermodynamic temperature	T
amount of substance	n
luminous intensity	I_{v}

All other physical quantities are called *derived quantities* and are regarded as having dimensions algebraically derived from the seven base quantities by multiplication and division.

Example dimension of (energy) = dimension of (mass × length^2 × time^{-2})

The physical quantity *amount of substance* is of special importance to chemists. Amount of substance is proportional to the number of specified elementary entities of that substance, the proportionality factor being the same for all substances; its reciprocal is the *Avogadro constant* (see sections 2.10, p.41, and 3.2, p.64, and chapter 5). The SI unit of amount of substance is the mole, defined in chapter 3 below. The physical quantity 'amount of substance' should no longer be called 'number of moles', just as the physical quantity 'mass' should not be called 'number of kilograms'. The name 'amount of substance' may often be usefully abbreviated to the single word 'amount', particularly in such phrases as 'amount concentration' (p.38), and 'amount of N_2' (see examples on p.42).

1.3 SYMBOLS FOR PHYSICAL QUANTITIES AND UNITS [4.a]

A clear distinction should be drawn between the names and symbols for physical quantities, and the names and symbols for units. Names and symbols for many physical quantities are given in chapter 2; the symbols given there are *recommendations*. If other symbols are used they should be clearly defined. Names and symbols for units are given in chapter 3; the symbols for units listed there are *mandatory*.

General rules for symbols for physical quantities

The symbol for a physical quantity should generally be a single letter of the Latin or Greek alphabet (see p.124)[1]. Capital and lower case letters may both be used. The letter should be printed in italic (sloping) type. When no italic font is available the distinction may be made by underlining symbols for physical quantities in accord with standard printers' practice. When necessary the symbol may be modified by subscripts and/or superscripts of specified meaning. Subscripts and superscripts that are themselves symbols for physical quantities or numbers should be printed in italic type; other subscripts and superscripts should be printed in roman (upright) type.

Examples C_p for heat capacity at constant pressure
x_i for mole fraction of the *i*th species
but C_{B} for heat capacity of substance B
E_{k} for kinetic energy
μ_{r} for relative permeability
$\Delta_{\mathrm{r}}H^{\ominus}$ for standard reaction enthalpy
V_{m} for molar volume

The meaning of symbols for physical quantities may be further qualified by the use of one or more subscripts, or by information contained in round brackets.

Examples $\Delta_{\mathrm{f}}S^{\ominus}(\mathrm{HgCl_2}, \mathrm{cr}, 25\,°\mathrm{C}) = -154.3\ \mathrm{J\ K^{-1}\ mol^{-1}}$
$\mu_i = (\partial G/\partial n_i)_{T,p,n_{j\neq i}}$

Vectors and matrices should be printed in bold face italic type, e.g. $\boldsymbol{A}$, $\boldsymbol{a}$. Matrices and tensors are sometimes printed in bold face sans-serif type, e.g. **S**, **T**. Vectors may alternatively be characterized by an arrow, $\vec{A}$, $\vec{a}$, and second rank tensors by a double arrow, $\vec{\vec{S}}$, $\vec{\vec{T}}$.

General rules for symbols for units

Symbols for units should be printed in roman (upright) type. They should remain unaltered in the plural, and should not be followed by a full stop except at the end of a sentence.

Example $r = 10$ cm, not cm. or cms.

Symbols for units should be printed in lower-case letters, unless they are derived from a personal name when they should begin with a capital letter. (An exception is the symbol for the litre which may be either L or l, i.e. either capital or lower case.)

Examples m (metre), s (second), but J (joule), Hz (hertz)

---

(1) An exception is made for certain dimensionless quantities used in the study of transport processes for which the internationally agreed symbols consist of two letters (see section 2.15).

Example Reynolds number, *Re*

When such symbols appear as factors in a product, they should be separated from other symbols by a space, multiplication sign, or brackets.

Decimal multiples and submultiples of units may be indicated by the use of prefixes as defined in section 3.6 below.

Examples nm (nanometre), kHz (kilohertz), Mg (megagram)

1.4 USE OF THE WORDS 'SPECIFIC' AND 'MOLAR' IN THE NAMES OF PHYSICAL QUANTITIES

The adjective *specific* before the name of an extensive quantity is often used to mean *divided by mass.* When the symbol for the extensive quantity is a capital letter, the symbol used for the specific quantity is often the corresponding lower case letter.

Examples volume, V
specific volume, $v = V/m = 1/\rho$ (where ρ is mass density)
heat capacity at constant pressure, C_p
specific heat capacity at constant pressure, $c_p = C_p/m$

The adjective *molar* before the name of an extensive quantity generally means *divided by amount of substance.* The subscript m on the symbol for the extensive quantity denotes the corresponding molar quantity.

Examples	volume, V	molar volume, $V_{\mathrm{m}} = V/n$ (p.37)
	enthalpy, H	molar enthalpy, $H_{\mathrm{m}} = H/n$

It is sometimes convenient to divide all extensive quantities by amount of substance, so that all quantities become intensive; the subscript m may then be omitted if this convention is stated and there is no risk of ambiguity. (See also the symbols recommended for partial molar quantities in section 2.11, p.44, and 'Examples of the use of these symbols', p.46.)

There are a few cases where the adjective *molar* has a different meaning, namely *divided by amount-of-substance concentration.*

Examples absorption coefficient, a
molar absorption coefficient, $\varepsilon = a/c$ (p.30)
conductivity, κ
molar conductivity, $\Lambda = \kappa/c$ (p.53)

1.5 PRODUCTS AND QUOTIENTS OF PHYSICAL QUANTITIES AND UNITS

Products of physical quantities may be written in any of the ways

$$ab \quad \text{or} \quad a \cdot b \quad \text{or} \quad a \times b$$

and similarly quotients may be written

$$a/b \quad \text{or} \quad \frac{a}{b} \quad \text{or} \quad ab^{-1}$$

Examples $F = ma$, $p = nRT/V$

Not more than one solidus (/) should be used in the same expression unless brackets are used to eliminate ambiguity.

Example $(a/b)/c$, but never $a/b/c$

In evaluating combinations of many factors, multiplication takes precedence over division in the sense that a/bc should be interpreted as $a/(bc)$ rather than $(a/b)c$; however, in complex expressions it is desirable to use brackets to eliminate any ambiguity.

Products and quotients of units may be written in a similar way, except that when a product of units is written without any multiplication sign one space should be left between the unit symbols.

Example $\mathrm{N = m\ kg\ s^{-2}}$, but not $\mathrm{mkgs^{-2}}$

When two or more units occur after single solidus, they are all to be read as in the denominator, but in complicated cases it is desirable to use parentheses to avoid ambiguity.[2]

Example $\mathrm{J\ K^{-1}\ mol^{-1}}$, or J/K mol, or J/(K mol).

---

(2) CGPM [2] specify that parentheses should always be used in this situation.

1.6 THE USE OF ABBREVIATIONS

Abbreviations and acronyms (words formed from the initial letters of groups of words that are frequently repeated) should be used sparingly. Unless they are well established (e.g. NMR, DNA) they should always be defined once in any paper, and they should generally be avoided in titles and abstracts. Abbreviations used to denote physical quantities should if possible be replaced by the recommended symbol for the quantity (e.g. E_i rather than I.P. for ionization energy, see p.19; ρ rather than dens. for mass density, see p.14). For further recommendations concerning abbreviations see [7].

2
Tables of physical quantities

The following tables contain the internationally recommended names and symbols for the physical quantities most likely to be used by chemists. Further quantities and symbols may be found in recommendations by IUPAP [3] and ISO [4].

Although authors are free to choose any symbols they wish for the quantities they discuss, provided that they define their notation and conform to the general rules indicated in chapter 1, it is clearly an aid to scientific communication if we all generally follow a standard notation. The symbols below have been chosen to conform with current usage and to minimize conflict so far as possible. Small variations from the recommended symbols may often be desirable in particular situations, perhaps by adding or modifying subscripts and/or superscripts, or by the alternative use of upper or lower case. Within a limited subject area it may also be possible to simplify notation, for example by omitting qualifying subscripts or superscripts, without introducing ambiguity. The notation adopted should in any case always be defined. Major deviations from the recommended symbols should be particularly carefully defined.

The tables are arranged by subject. The five columns in each table give the name of the quantity, the recommended symbol(s), a brief definition, the symbol for the coherent SI unit (without multiple or submultiple prefixes, see p.68), and footnote references. When two or more symbols are recommended, commas are used to separate symbols that are equally acceptable, and symbols of second choice are put in parentheses. The definitions are given primarily for identification purposes and are not necessarily complete; they should be regarded as useful relations rather than formal definitions. For dimensionless quantities a 1 is entered in the SI unit column. Further information is added in footnotes, and in text inserts between the tables, as appropriate.

2.1 SPACE AND TIME

The names and symbols recommended here are in agreement with those recommended by IUPAP [3] and ISO [4.b,c].

Name	*Symbol*	*Definition*	*SI unit*	*Notes*
cartesian space coordinates	x, y, z		m	
spherical polar coordinates	r, θ, ϕ		m, 1, 1	
generalized coordinate	q, q_i		(varies)	
position vector	$\boldsymbol{r}$	$\boldsymbol{r}=x\boldsymbol{i}+y\boldsymbol{j}+z\boldsymbol{k}$	m	
length	l		m	
special symbols:				
height	h			
breadth	b			
thickness	d, δ			
distance	d			
radius	r			
diameter	d			
path length	s			
length of arc	s			
area	A, A_s, S		m^2	1
volume	$V, (v)$		m^3	
plane angle	$\alpha, \beta, \gamma, \theta, \phi \ldots$	$\alpha=s/r$	rad, 1	2
solid angle	ω, Ω	$\omega=A/r^2$	sr, 1	2
time	t		s	
period	T	$T=t/N$	s	
frequency	ν, f	$\nu=1/T$	Hz	
circular frequency, angular frequency	ω	$\omega=2\pi\nu$	rad s^{-1}, s^{-1}	2, 3
characteristic time interval, relaxation time, time constant	τ, T	$\tau=\|\mathrm{d}t/\mathrm{d}\ln x\|$	s	
angular velocity	ω	$\omega=\mathrm{d}\phi/\mathrm{d}t$	rad s^{-1}, s^{-1}	2, 4
velocity	$\boldsymbol{v}, \boldsymbol{u}, \boldsymbol{w}, \boldsymbol{c}, \dot{\boldsymbol{r}}$	$\boldsymbol{v}=\mathrm{d}\boldsymbol{r}/\mathrm{d}t$	m s^{-1}	
speed	v, u, w, c	$v=\|\boldsymbol{v}\|$	m s^{-1}	5
acceleration	$\boldsymbol{a}, (g)$	$\boldsymbol{a}=\mathrm{d}\boldsymbol{v}/\mathrm{d}t$	m s^{-2}	6

(1) An infinitesimal area may be regarded as a vector $\mathrm{d}\boldsymbol{A}$ perpendicular to the plane. The symbol A_s may be used when necessary to avoid confusion with A for Helmholtz energy.
(2) The units radian (rad) and steradian (sr), for plane angle and solid angle respectively, are described as 'SI supplementary units' [2]. Since they are of dimension 1 (i.e. dimensionless), they may be included if appropriate, or they may be omitted if clarity is not lost thereby, in expressions for derived SI units.
(3) The unit Hz is not to be used for circular frequency.
(4) Angular velocity can be treated as a vector.
(5) For the speeds of light and sound the symbol c is customary.
(6) For acceleration of free fall the symbol g is used.

2.2 CLASSICAL MECHANICS

The names and symbols recommended here are in agreement with those recommended by IUPAP [3] and ISO [4.d]. Additional quantities and symbols used in acoustics can be found in [3 and 4.h].

Name	*Symbol*	*Definition*	*SI unit*	*Notes*
mass	m		kg	
reduced mass	μ	$\mu = m_1 m_2/(m_1 + m_2)$	kg	
density, mass density	ρ	$\rho = m/V$	kg m^{-3}	
relative density	d	$d = \rho/\rho^{\ominus}$	1	1
surface density	ρ_A, ρ_S	$\rho_A = m/A$	kg m^{-2}	
specific volume	v	$v = V/m = 1/\rho$	m^3 kg^{-1}	
momentum	$\boldsymbol{p}$	$\boldsymbol{p} = m\boldsymbol{v}$	kg m s^{-1}	
angular momentum, action	$\boldsymbol{L}$	$\boldsymbol{L} = \boldsymbol{r} \times \boldsymbol{p}$	J s	2
moment of inertia	I, J	$I = \sum m_i r_i^2$	kg m^2	3
force	$\boldsymbol{F}$	$\boldsymbol{F} = \mathrm{d}\boldsymbol{p}/\mathrm{d}t = m\boldsymbol{a}$	N	
torque, moment of a force	$\boldsymbol{T}$, ($\boldsymbol{M}$)	$\boldsymbol{T} = \boldsymbol{r} \times \boldsymbol{F}$	N m	
energy	E		J	
potential energy	E_p, V, Φ	$E_\mathrm{p} = -\int \boldsymbol{F} \cdot \mathrm{d}\boldsymbol{s}$	J	
kinetic energy	E_k, T, K	$E_\mathrm{k} = \frac{1}{2} m v^2$	J	
work	W, w	$W = \int \boldsymbol{F} \cdot \mathrm{d}\boldsymbol{s}$	J	
Hamilton function	H	$H(q, p) = T(q, p) + V(q)$	J	
Lagrange function	L	$L(q, \dot{q}) = T(q, \dot{q}) - V(q)$	J	
pressure	p, P	$p = F/A$	Pa, N m^{-2}	
surface tension	γ, σ	$\gamma = \mathrm{d}W/\mathrm{d}A$	N m^{-1}, J m^{-2}	
weight	G, (W, P)	$G = mg$	N	
gravitational constant	G	$F = G m_1 m_2/r^2$	N m^2 kg^{-2}	
normal stress	σ	$\sigma = F/A$	Pa	
shear stress	τ	$\tau = F/A$	Pa	
linear strain, relative elongation	ε, e	$\varepsilon = \Delta l/l$	1	
modulus of elasticity, Young's modulus	E	$E = \sigma/\varepsilon$	Pa	
shear strain	γ	$\gamma = \Delta x/d$	1	
shear modulus	G	$G = \tau/\gamma$	Pa	
volume strain, bulk strain	θ	$\theta = \Delta V/V_0$	1	
bulk modulus, compression modulus	K	$K = -V_0(\mathrm{d}p/\mathrm{d}V)$	Pa	

(1) Usually $\rho^{\ominus} = \rho(H_2O, 4\ ^\circ C)$.

(2) Other symbols are customary in atomic and molecular spectroscopy; see section 2.6.

(3) In general I is a tensor quantity: $I_{\alpha\alpha} = \Sigma\, m_i(\beta_i^2 + \gamma_i^2)$, and $I_{\alpha\beta} = -\Sigma\, m_i \alpha_i \beta_i$ if $\alpha \neq \beta$, where α, β, γ is a permutation of x, y, z.

Name	*Symbol*	*Definition*	*SI unit*	*Notes*
viscosity, dynamic viscosity	η, μ	$\tau_{x,z}=\eta(\mathrm{d}v_x/\mathrm{d}z)$	Pa s	
fluidity	ϕ	$\phi=1/\eta$	$\mathrm{m\ kg^{-1}\ s}$	
kinematic viscosity	ν	$\nu=\eta/\rho$	$\mathrm{m^2\ s^{-1}}$	
friction coefficient	μ, (f)	$F_{\mathrm{frict}}=\mu F_{\mathrm{norm}}$	1	
power	P	$P=\mathrm{d}W/\mathrm{d}t$	W	
sound energy flux	P, P_{a}	$P=\mathrm{d}E/\mathrm{d}t$	W	
acoustic factors,				
reflection factor	ρ	$\rho=P_{\mathrm{r}}/P_0$	1	4
acoustic absorption factor	α_{a}, (α)	$\alpha_{\mathrm{a}}=1-\rho$	1	5
transmission factor	τ	$\tau=P_{\mathrm{tr}}/P_0$	1	4
dissipation factor	δ	$\delta=\alpha_{\mathrm{a}}-\tau$	1	

(4) P_0 is the incident sound energy flux, P_{r} the reflected flux and P_{tr} the transmitted flux.

(5) This definition is special to acoustics and is different from the usage in radiation, where the absorption factor corresponds to the acoustic dissipation factor.

2.3 ELECTRICITY AND MAGNETISM

The names and symbols recommended here are in agreement with those recommended by IUPAP [3] and ISO [4.f].

Name	*Symbol*	*Definition*	*SI unit*	*Notes*
quantity of electricity, electric charge	Q		C	
charge density	ρ	$\rho = Q/V$	$\mathrm{C\,m^{-3}}$	
surface charge density	σ	$\sigma = Q/A$	$\mathrm{C\,m^{-2}}$	
electric potential	V, ϕ	$V = \mathrm{d}W/\mathrm{d}Q$	$\mathrm{V, J\,C^{-1}}$	
electric potential difference	$U, \Delta V, \Delta\phi$	$U = V_2 - V_1$	V	
electromotive force	E	$E = \int (\boldsymbol{F}/Q)\cdot \mathrm{d}\boldsymbol{s}$	V	
electric field strength	$\boldsymbol{E}$	$\boldsymbol{E} = \boldsymbol{F}/Q = -\mathrm{grad}\, V$	$\mathrm{V\,m^{-1}}$	
electric flux	Ψ	$\Psi = \int \boldsymbol{D}\cdot \mathrm{d}\boldsymbol{A}$	C	1
electric displacement	$\boldsymbol{D}$	$\boldsymbol{D} = \varepsilon \boldsymbol{E}$	$\mathrm{C\,m^{-2}}$	
capacitance	C	$C = Q/U$	$\mathrm{F, C\,V^{-1}}$	
permittivity	ε	$\boldsymbol{D} = \varepsilon \boldsymbol{E}$	$\mathrm{F\,m^{-1}}$	
permittivity of vacuum	ε_0	$\varepsilon_0 = \mu_0^{-1} c_0^{-2}$	$\mathrm{F\,m^{-1}}$	
relative permittivity	ε_r	$\varepsilon_r = \varepsilon/\varepsilon_0$	1	2
dielectric polarization (dipole moment per volume)	$\boldsymbol{P}$	$\boldsymbol{P} = \boldsymbol{D} - \varepsilon_0 \boldsymbol{E}$	$\mathrm{C\,m^{-2}}$	
electric susceptibility	χ_e	$\chi_e = \varepsilon_r - 1$	1	
electric dipole moment	$\boldsymbol{p}, \boldsymbol{\mu}$	$\boldsymbol{p} = Q\boldsymbol{r}$	C m	3
electric current	I	$I = \mathrm{d}Q/\mathrm{d}t$	A	
electric current density	$\boldsymbol{j}, \boldsymbol{J}$	$I = \int \boldsymbol{j}\cdot \mathrm{d}\boldsymbol{A}$	$\mathrm{A\,m^{-2}}$	1
magnetic flux density, magnetic induction	$\boldsymbol{B}$	$\boldsymbol{F} = Q\boldsymbol{v} \times \boldsymbol{B}$	T	4
magnetic flux	Φ	$\Phi = \int \boldsymbol{B}\cdot \mathrm{d}\boldsymbol{A}$	Wb	1
magnetic field strength	$\boldsymbol{H}$	$\boldsymbol{B} = \mu \boldsymbol{H}$	$\mathrm{A\,m^{-1}}$	
permeability	μ	$\boldsymbol{B} = \mu \boldsymbol{H}$	$\mathrm{N\,A^{-2}, H\,m^{-1}}$	
permeability of vacuum	μ_0		$\mathrm{H\,m^{-1}}$	
relative permeability	μ_r	$\mu_r = \mu/\mu_0$	1	
magnetization (magnetic dipole moment per volume)	$\boldsymbol{M}$	$\boldsymbol{M} = \boldsymbol{B}/\mu_0 - \boldsymbol{H}$	$\mathrm{A\,m^{-1}}$	
magnetic susceptibility	$\chi, \kappa, (\chi_m)$	$\chi = \mu_r - 1$	1	5
molar magnetic susceptibility	χ_m	$\chi_m = V_m \chi$	$\mathrm{m^3\,mol^{-1}}$	

(1) $\mathrm{d}\boldsymbol{A}$ is a vector element of area.
(2) This quantity was formerly called dielectric constant.
(3) When a dipole is composed of two point charges, the direction of the dipole vector is taken to be from the negative to the positive charge.
(4) This quantity is sometimes loosely called magnetic field.
(5) The symbol χ_m is sometimes used for magnetic susceptibility, but it should be reserved for molar magnetic susceptibility.

Name	Symbol	Definition	SI unit	Notes
magnetic dipole moment	$\boldsymbol{m}, \boldsymbol{\mu}$	$E_p = -\boldsymbol{m} \cdot \boldsymbol{B}$	$\mathrm{A\,m^2}$, $\mathrm{J\,T^{-1}}$	
electrical resistance	R	$R = U/I$	Ω	6
conductance	G	$G = 1/R$	S	6
loss angle	δ	$\delta = (\pi/2) + \phi_I - \phi_U$	1, rad	7
reactance	X	$X = (U/I)\sin\delta$	Ω	
impedance (complex impedance)	Z	$Z = R + \mathrm{i}X$	Ω	
admittance (complex admittance)	Y	$Y = 1/Z$	S	
susceptance	B	$Y = G + \mathrm{i}B$	S	
resistivity	ρ	$\rho = E/j$	$\Omega\,\mathrm{m}$	8
conductivity	κ, γ, σ	$\kappa = 1/\rho$	$\mathrm{S\,m^{-1}}$	8
self-inductance	L	$E = -L(\mathrm{d}I/\mathrm{d}t)$	H	
mutual inductance	M, L_{12}	$E_1 = L_{12}(\mathrm{d}I_2/\mathrm{d}t)$	H	
magnetic vector potential	$\boldsymbol{A}$	$\boldsymbol{B} = \nabla \times \boldsymbol{A}$	$\mathrm{Wb\,m^{-1}}$	
Poynting vector	$\boldsymbol{S}$	$\boldsymbol{S} = \boldsymbol{E} \times \boldsymbol{H}$	$\mathrm{W\,m^{-2}}$	

(6) In a material with reactance $R = (U/I)\cos\delta$, and $G = R/(R^2 + X^2)$.
(7) ϕ_I and ϕ_U are the phases of current and potential difference.
(8) These quantities are tensors in anisotropic materials.

2.4 QUANTUM MECHANICS

The names and symbols recommended here are in agreement with those recommended by IUPAP [3].

Name	*Symbol*	*Definition*	*SI unit*	*Notes*
momentum operator	$\hat{p}$	$\hat{p} = -\mathrm{i}\hbar\nabla$	$\mathrm{m^{-1}\,J\,s}$	
kinetic energy operator	$\hat{T}$	$\hat{T} = -(\hbar^2/2m)\nabla^2$	J	
hamiltonian operator	$\hat{H}$	$\hat{H} = \hat{T} + V$	J	
wavefunction, state function	Ψ, ψ, ϕ	$\hat{H}\psi = E\psi$	$(\mathrm{m^{-3/2}})$	1, 2
probability density	P	$P = \psi^*\psi$	$(\mathrm{m^{-3}})$	2, 3
charge density of electrons	ρ	$\rho = -eP$	$(\mathrm{C\,m^{-3}})$	2, 4
probability current density	$\boldsymbol{S}$	$\boldsymbol{S} = -\mathrm{i}\hbar(\psi^*\nabla\psi - \psi\nabla\psi^*)/2m_e$	$(\mathrm{m^{-2}\,s^{-1}})$	2
electric current density of electrons	$\boldsymbol{j}$	$\boldsymbol{j} = -e\boldsymbol{S}$	$(\mathrm{A\,m^{-2}})$	2, 4
matrix element of operator $\hat{A}$	$A_{ij}, \langle i\|\hat{A}\|j\rangle$	$A_{ij} = \int\psi_i^*\hat{A}\psi_j\,\mathrm{d}\tau$	(varies)	5
expectation value of operator $\hat{A}$	$\langle A\rangle, \bar{A}$	$\langle A\rangle = \int\psi^*\hat{A}\psi\,\mathrm{d}\tau$	(varies)	5
hermitian conjugate of $\hat{A}$	$\hat{A}^\dagger$	$(\hat{A}^\dagger)_{ij} = (A_{ji})^*$	(varies)	5
commutator of $\hat{A}$ and $\hat{B}$	$[\hat{A}, \hat{B}], [\hat{A}, \hat{B}]_-$	$[\hat{A}, \hat{B}] = \hat{A}\hat{B} - \hat{B}\hat{A}$	(varies)	6
anticommutator	$[\hat{A}, \hat{B}]_+$	$[\hat{A}, \hat{B}]_+ = \hat{A}\hat{B} + \hat{B}\hat{A}$	(varies)	6
spin wavefunction	$\alpha; \beta$		1	
coulomb integral	H_{AA}	$H_{AA} = \int\psi_A^*\hat{H}\psi_A\,\mathrm{d}\tau$	J	7
resonance integral	H_{AB}	$H_{AB} = \int\psi_A^*\hat{H}\psi_B\,\mathrm{d}\tau$	J	7
overlap integral	S_{AB}	$S_{AB} = \int\psi_A^*\psi_B\,\mathrm{d}\tau$	1	7

(1) Capital and lower-case psi are often used for the time-dependent function $\Psi(x, t)$ and the amplitude function $\psi(x)$ respectively. Thus for a stationary state $\Psi(x, t) = \psi(x)\exp(-\mathrm{i}Et/\hbar)$.
(2) For the normalized wavefunction of a single particle in three-dimensional space the appropriate SI unit is given in parentheses. Results in quantum mechanics are, however, usually expressed in terms of atomic units (see section 3.8, 7.3, and reference [8]).
(3) ψ^* is the complex conjugate of ψ.
(4) $-e$ is the charge of an electron.
(5) The unit is the same as for the physical quantity that the operator represents.
(6) The unit is the same as for the product of the corresponding physical quantities A and B.
(7) $\hat{H}$ is an effective hamiltonian for a single electron, and A and B label the orbitals. In the special context of Hückel molecular orbital (MO) theory, H_{AA} and H_{AB} are often denoted α and β respectively.

2.5 ATOMS AND MOLECULES

The names and symbols recommended here are in agreement with those recommended by IUPAP [3] and ISO [4.j]. Additional quantities and symbols used in atomic, nuclear and plasma physics can be found in [3 and 4.k].

Name	*Symbol*	*Definition*	*SI unit*	*Notes*
nucleon number, mass number	A		1	
proton number, atomic number	Z		1	
neutron number	N	$N = A - Z$	1	
electron rest mass	m_e		kg	1, 2
mass of atom, atomic mass	m_a, m		kg	
atomic mass constant	m_u	$m_u = m_a(^{12}C)/12$	kg	1, 3
mass excess	Δ	$\Delta = m_a - Am_u$	kg	
elementary charge, proton charge	e		C	2
Planck constant	h		J s	
Planck constant/2π	$\hbar$	$\hbar = h/2\pi$	J s	2
Bohr radius	a_0	$a_0 = 4\pi\varepsilon_0\hbar^2/m_e e^2$	m	2
Hartree energy	E_h	$E_h = \hbar^2/m_e {a_0}^2$	J	2
Rydberg constant	R_∞	$R_\infty = E_h/2hc$	m^{-1}	
fine structure constant	α	$\alpha = e^2/4\pi\varepsilon_0\hbar c$	1	
ionization energy	E_i		J	
electron affinity	E_{ea}		J	
dissociation energy	E_d, D		J	
from the ground state	D_0		J	4
from the potential minimum	D_e		J	4
principal quantum number (H atom)	n	$E = -hcR/n^2$	1	
angular momentum quantum numbers	see under Spectroscopy, section 2.6			
magnetic dipole moment of a molecule	$\boldsymbol{m}$, $\boldsymbol{\mu}$	$E_p = -\boldsymbol{m} \cdot \boldsymbol{B}$	$J\ T^{-1}$	5
magnetizability of a molecule	ξ	$\boldsymbol{m} = \xi \boldsymbol{B}$	$J\ T^{-2}$	

(1) Analogous symbols are used for other particles with subscripts: p for proton, n for neutron, a for atom, N for nucleus, etc.
(2) This quantity is also used as an atomic unit; see sections 3.8 and 7.3.
(3) m_u is equal to the unified atomic mass unit, with symbol u, i.e. $m_u = 1$ u (see section 3.7). In biochemistry the name dalton, with symbol Da, is used for the unified atomic mass unit, although the name and symbol have not been accepted by CGPM.
(4) The symbols D_0 and D_e are mainly used for diatomic dissociation energies.
(5) Magnetic moments of specific particles may be denoted by subscripts, e.g. μ_e, μ_p, μ_n for an electron, a proton, and a neutron. Tabulated values usually refer to the maximum expectation value of the z component. Values for stable nuclei are given in table 6.2.

Name	*Symbol*	*Definition*	*SI unit*	*Notes*
Bohr magneton	μ_B	$\mu_B = e\hbar/2m_e$	$J\ T^{-1}$	
nuclear magneton	μ_N	$\mu_N = (m_e/m_p)\mu_B$	$J\ T^{-1}$	
magnetogyric ratio (gyromagnetic ratio)	γ	$\gamma = \mu/L$	$C\ kg^{-1}$	
g factor	g		1	
Larmor circular frequency	ω_L	$\omega_L = (e/2m)B$	s^{-1}	
Larmor frequency	ν_L	$\nu_L = \omega_L/2\pi$	Hz	
longitudinal relaxation time	T_1		s	6
transverse relaxation time	T_2		s	6
electric dipole moment of a molecule	$\boldsymbol{p}, \boldsymbol{\mu}$	$E_p = -\boldsymbol{p}\cdot\boldsymbol{E}$	C m	
quadrupole moment of a molecule	$\boldsymbol{Q}; \boldsymbol{\Theta}$	$E_p = \frac{1}{2}\boldsymbol{Q}:\boldsymbol{V}'' = \frac{1}{3}\boldsymbol{\Theta}:\boldsymbol{V}''$	$C\ m^2$	7
quadrupole moment of a nucleus	eQ	$eQ = 2\langle\Theta_{zz}\rangle$	$C\ m^2$	8
electric field gradient tensor	$\boldsymbol{q}$	$q_{\alpha\beta} = -\partial^2 V/\partial\alpha\partial\beta$	$V\ m^{-2}$	
quadrupole interaction energy tensor	$\boldsymbol{\chi}$	$\chi_{\alpha\beta} = eQq_{\alpha\beta}$	J	9
electric polarizability of a molecule	α	p (induced) $= \alpha E$	$C\ m^2\ V^{-1}$	10
activity (of a radioactive substance)	A	$A = -\mathrm{d}N_B/\mathrm{d}t$	Bq	11
decay (rate) constant, disintegration (rate) constant	λ	$A = \lambda N_B$	s^{-1}	11

(6) These quantities are used in the context of saturation effects in spectroscopy, particularly spin-resonance spectroscopy (see p.24).

(7) The quadrupole moment of a molecule may be represented either by the tensor $\boldsymbol{Q}$, defined by an integral over the charge density ρ:

$$Q_{\alpha\beta} = \int r_\alpha r_\beta \rho \mathrm{d}V$$

where α and β denote x, y or z, or by the traceless tensor $\boldsymbol{\Theta}$ defined by

$$\Theta_{\alpha\beta} = (1/2)\int(3r_\alpha r_\beta - \delta_{\alpha\beta}r^2)\rho\mathrm{d}V$$
$$= (1/2)\,[3Q_{\alpha\beta} - \delta_{\alpha\beta}(Q_{xx} + Q_{yy} + Q_{zz})]$$

$\boldsymbol{V}''$ is the second derivative of the electric potential;

$$V_{\alpha\beta}'' = -q_{\alpha\beta} = \partial^2 V/\partial\alpha\partial\beta$$

(8) Nuclear quadrupole moments are conventionally defined in a different way from molecular quadrupole moments. Q is an area and e is the elementary charge. eQ is taken to be the maximum expectation value of the zz tensor element. The values of Q for some nuclei are listed in table 6.2.

(9) The nuclear quadrupole interaction energy tensor $\boldsymbol{\chi}$ is usually quoted in MHz, corresponding to the value of eQq/h, although the h is usually omitted.

(10) The molecular polarizability is a symmetric tensor with in general six independent components: $p_\alpha = \Sigma\,\alpha_{\alpha\beta}E_\beta$ (where α, β are indices for x, y and z).

(11) N_B is the number of radioactive atoms B.

Name	*Symbol*	*Definition*	*SI unit*	*Notes*
half life	$t_{\frac{1}{2}}, T_{\frac{1}{2}}$		s	
mean life	τ		s	
level width	Γ	$\Gamma = \hbar/\tau$	J	
disintegration energy	Q		J	
cross section (of a nuclear reaction)	σ		m^2	

2.6 SPECTROSCOPY

This section has been considerably extended compared with the previous edition [1.c] and with the corresponding section in the IUPAP document [3]. It is based on the recommendations of the ICSU Joint Commission for Spectroscopy [36, 37] and current practice in the field which is well represented in the books by Herzberg [38]. The IUPAC Commission on Molecular Structure and Spectroscopy has also published various recommendations which have been taken into account [9–15].

Name	*Symbol*	*Definition*	*SI unit*	*Notes*
total term	T	$T=E_{\text{tot}}/hc$	m^{-1}	1, 2
transition wavenumber	$\tilde{\nu}$, (ν)	$\tilde{\nu}=T'-T''$	m^{-1}	1
transition frequency	ν	$\nu=(E'-E'')/h$	Hz	
electronic term	T_{e}	$T_{\text{e}}=E_{\text{e}}/hc$	m^{-1}	1, 2
vibrational term	G	$G=E_{\text{vib}}/hc$	m^{-1}	1, 2
rotational term	F	$F=E_{\text{rot}}/hc$	m^{-1}	1, 2
spin orbit coupling constant	A	$T_{\text{s.o.}}=A\langle\hat{\boldsymbol{L}}\cdot\hat{\boldsymbol{S}}\rangle$	m^{-1}	1
principal moments of inertia	I_A; I_B; I_C	$I_A \leqslant I_B \leqslant I_C$	kg m^2	
rotational constants,				
in wavenumber	$\tilde{A}$; $\tilde{B}$; $\tilde{C}$	$\tilde{A}=h/8\pi^2 cI_A$	m^{-1}	1, 2
in frequency	A; B; C	$A=h/8\pi^2 I_A$	Hz	
inertial defect	Δ	$\Delta=I_C-I_A-I_B$	kg m^2	
asymmetry parameter	κ	$\kappa=\dfrac{(2B-A-C)}{(A-C)}$	1	3
centrifugal distortion constants,				
S reduction	D_J; D_{JK}; D_K; d_1; d_2		m^{-1}	4
A reduction	Δ_J; Δ_{JK}; Δ_K; δ_J; δ_K		m^{-1}	4
harmonic vibration wavenumber	ω_{e}; ω_r		m^{-1}	5
vibrational anharmonicity constant	$\omega_{\text{e}}x_{\text{e}}$; x_{rs}; $g_{tt'}$		m^{-1}	5
vibrational quantum numbers	v_r; l_t		1	5

(1) In spectroscopy the unit cm^{-1} is almost always used for wavenumber, and term values and wavenumbers always refer to the reciprocal wavelength of the equivalent radiation in vacuum. The symbol c in the definition E/hc refers to the speed of light in vacuum.

(2) Term values and rotational constants are sometimes defined in wavenumber units (e.g. $T=E/hc$), and sometimes in frequency units (e.g. $T=E/h$). When the symbol is otherwise the same, it is convenient to distinguish wavenumber quantities with a tilde (e.g. $\tilde{\nu}$, $\tilde{T}$, $\tilde{A}$, $\tilde{B}$, $\tilde{C}$ for quantities defined in wavenumber units), although this is not a universal practice.

(3) The Wang asymmetry parameters are also used: for a near prolate top $b_{\text{p}}=(C-B)/(2A-B-C)$, and for a near oblate top $b_{\text{o}}=(A-B)/(2C-A-B)$.

(4) S and A stand for the symmetric and asymmetric reductions of the rotational hamiltonian respectively; see [39] for more details on the various possible representations of the centrifugal distortion constants.

(5) For a diatomic: $G(v)=\omega_{\text{e}}(v+\frac{1}{2})-\omega_{\text{e}}x_{\text{e}}(v+\frac{1}{2})^2+\ldots$ For a polyatomic molecule the $3N-6$ vibrational modes ($3N-5$ if linear) are labelled by the indices $r, s, t, \ldots$, or $i, j, k, \ldots$. The index r is usually assigned in descending wavenumber order, symmetry species by symmetry species. The index t is kept for degenerate modes. The vibrational term formula is

$$G(v)=\sum_r \omega_r(v_r+d_r/2)+\sum_{r\leqslant s} x_{rs}(v_r+d_r/2)(v_s+d_s/2)+\sum_{t\leqslant t'} g_{tt'}l_t l_{t'}+\ldots$$

Name	Symbol	Definition	SI unit	Notes
Coriolis zeta constant	ζ_{rs}^{α}		1	
angular momentum quantum numbers	see additional information below			
degeneracy, statistical weight	g, d, β		1	6
electric dipole moment of a molecule	$\boldsymbol{p}$, $\boldsymbol{\mu}$	$E_{\mathrm{p}} = -\boldsymbol{p} \cdot \boldsymbol{E}$	C m	7
transition dipole moment of a molecule	$\boldsymbol{M}$, $\boldsymbol{R}$	$\boldsymbol{M} = \int \psi' \boldsymbol{p} \psi'' \mathrm{d}\tau$	C m	7
molecular geometry, interatomic distances,				8, 9
equilibrium distance	r_{e}		m	
zero-point average distance	r_{z}		m	
ground state distance	r_0		m	
substitution structure distance	r_{s}		m	
vibrational coordinates,				8
internal coordinates	R_i, r_i, θ_j, etc.		(varies)	
symmetry coordinates	S_i		(varies)	
normal coordinates				
mass adjusted	Q_r		$\mathrm{kg}^{\frac{1}{2}}\,\mathrm{m}$	
dimensionless	q_r		1	
vibrational force constants,				10
diatomic	f, (k)	$f = \partial^2 V / \partial r^2$	$\mathrm{J\,m^{-2}}$	
polyatomic,				
internal coordinates	f_{ij}	$f_{ij} = \partial^2 V / \partial r_i \partial r_j$	(varies)	
symmetry coordinates	F_{ij}	$F_{ij} = \partial^2 V / \partial S_i \partial S_j$	(varies)	
dimensionless normal coordinates	$\phi_{rst\ldots}$, $k_{rst\ldots}$		$\mathrm{m^{-1}}$	11
nuclear magnetic resonance (NMR),				
magnetogyric ratio	γ	$\gamma = \mu / I\hbar$	$\mathrm{C\,kg^{-1}}$	
shielding constant	σ_{A}	$B_{\mathrm{A}} = (1 - \sigma_{\mathrm{A}})B$	1	12

(6) d is usually used for vibrational degeneracy, and β for nuclear spin degeneracy.
(7) Molecular dipole moments are often expressed in the non-SI unit debye, where $\mathrm{D} \approx 3.335\,64 \times 10^{-30}$ C m.
(8) Interatomic (internuclear) distances and vibrational displacements are often expressed in the non-SI unit ångström, where $\mathrm{Å} = 10^{-10}\,\mathrm{m} = 0.1\,\mathrm{nm} = 100\,\mathrm{pm}$.
(9) The various slightly different ways of representing interatomic distances, distinguished by subscripts, involve different vibrational averaging contributions; they are discussed in [40], where the geometrical structure of many free molecules is listed. Only the equilibrium distance r_{e} is isotopically invariant. The effective distance parameter r_0 is estimated from the rotational constants for the ground vibrational state and has only approximate physical significance for polyatomic molecules.
(10) Force constants are often expressed in $\mathrm{mdyn\,Å^{-1}} = \mathrm{aJ\,Å^{-2}}$ for stretching coordinates, $\mathrm{mdyn\,Å} = \mathrm{aJ}$ for bending coordinates, and $\mathrm{mdyn} = \mathrm{aJ\,Å^{-1}}$ for stretch–bend interactions. See [16] for further details on definitions and notation for force constants.
(11) The force constants in dimensionless normal coordinates are usually defined in wavenumber units by the equation $V/hc = \Sigma \phi_{rst\ldots} q_r q_s q_t \ldots$, where the summation over the normal coordinate indices $r, s, t, \ldots$ is unrestricted.
(12) σ_{A} and B_{A} denote the shielding constant and the local magnetic field at nucleus A.

Name	*Symbol*	*Definition*	*SI unit*	*Notes*
chemical shift, δ scale	δ	$\delta = 10^6(\nu - \nu_0)/\nu_0$	1	13
(indirect) spin–spin coupling constant	J_{AB}	$\hat{H}/h = J_{AB}\hat{\boldsymbol{I}}_A \cdot \hat{\boldsymbol{I}}_B$	Hz	14
direct (dipolar) coupling constant	D_{AB}		Hz	15
longitudinal relaxation time	T_1		s	16
transverse relaxation time	T_2		s	16
electron spin resonance, electron paramagnetic resonance (ESR, EPR),				
magnetogyric ratio	γ	$\gamma = \mu/s\hbar$	C kg^{-1}	
g factor	g	$h\nu = g\mu_B B$	1	
hyperfine coupling constant,				
in liquids	a, A	$\hat{H}_{hfs}/h = a\hat{\boldsymbol{S}}\cdot\hat{\boldsymbol{I}}$	Hz	17
in solids	$\boldsymbol{T}$	$\hat{H}_{hfs}/h = \hat{\boldsymbol{S}}\cdot\boldsymbol{T}\cdot\hat{\boldsymbol{I}}$	Hz	17

(13) ν_0 is the resonance frequency of a reference molecule, usually tetramethylsilane for proton and for ^{13}C resonance spectra [11]. In some of the older literature proton chemical shifts are expressed on the τ scale, where $\tau = 10 - \delta$, but this is no longer used.
(14) $\hat{H}$ in the definition is the spin–spin coupling hamiltonian between nuclei A and B.
(15) Direct dipolar coupling occurs in solids; the definition of the coupling constant is $D_{AB} = (\mu_0/4\pi)r_{AB}^{-3}\gamma_A\gamma_B(\hbar/2\pi)$.
(16) The longitudinal relaxation time is associated with spin–lattice relaxation, and the transverse relaxation time with spin–spin relaxation. The definitions are

$$\mathrm{d}M_z/\mathrm{d}t = -(M_z - M_{z,e})/T_1,$$

and

$$\mathrm{d}M_x/\mathrm{d}t = -M_x/T_2,$$

where M_z and M_x are the components of magnetization parallel and perpendicular to the static field B, and $M_{z,e}$ is the equilibrium value of M_z.
(17) $\hat{H}_{hfs}$ is the hyperfine coupling hamiltonian. The coupling constants a are usually quoted in MHz, but they are sometimes quoted in magnetic induction units (G or T) obtained by dividing by the conversion factor $g\mu_B/h$, which has the SI unit Hz/T; $g_e\mu_B/h \approx 2.8025$ MHz T^{-1}, where g_e is the g factor for a free electron. In liquids the hyperfine coupling is isotropic, and the coupling constant is a scalar a. In solids the coupling is anisotropic, and the coupling constant is a 3×3 tensor $\boldsymbol{T}$.

Symbols for angular momentum operators and quantum numbers

In the following table, all of the symbols denote (*angular momentum*)/$\hbar$, and are dimensionless. (Although this is a universal practice for the quantum numbers, some authors use the operator symbols to denote *angular momentum*, in which case the operators would have SI units: J s.) The column heading 'Z-axis' denotes the space-fixed component, and the heading 'z-axis' denotes the molecule fixed component along the symmetry axis (linear or symmetric top molecules), or the axis of quantization.

Angular momentum[1]	Operator symbol	Quantum number symbol Total	Z-axis	z-axis	Notes
electron orbital	$\hat{\boldsymbol{L}}$	L	M_L	Λ	2
one electron only	$\hat{\boldsymbol{l}}$	l	m_l	λ	2
electron spin	$\hat{\boldsymbol{S}}$	S	M_S	Σ	
one electron only	$\hat{\boldsymbol{s}}$	s	m_s	σ	
electron orbital + spin	$\hat{\boldsymbol{L}}+\hat{\boldsymbol{S}}$			$\Omega=\Lambda+\Sigma$	2
nuclear orbital (rotational)	$\hat{\boldsymbol{R}}$	R		K_R, k_R	
nuclear spin	$\hat{\boldsymbol{I}}$	I	M_I		
internal vibrational					
spherical top	$\hat{\boldsymbol{l}}$	$l(l\zeta)$		K_l	
other	$\hat{\boldsymbol{j}}, \hat{\boldsymbol{\pi}}$			$l(l\zeta)$	2
sum of $R+L(+j)$	$\hat{\boldsymbol{N}}$	N		K, k	2
sum of $N+S$	$\hat{\boldsymbol{J}}$	J	M_J	K, k	2, 4
sum of $J+I$	$\hat{\boldsymbol{F}}$	F	M_F		

(1) In all cases the vector operator and its components are related to the quantum numbers by eigenvalue equations analogous to:

$$\hat{J}^2\psi=J(J+1)\psi,\ \hat{J}_Z\psi=M_J\psi,\text{ and }\hat{J}_z\psi=K\psi,$$

where the component quantum numbers M_J and K take integral or half-odd values in the range $-J\leqslant M_J\leqslant +J$, $-J\leqslant K\leqslant +J$. (If the operator symbols are taken to represent *angular momentum*, rather than (*angular momentum*)/$\hbar$, the eigenvalue equations should read $\hat{J}^2\psi=J(J+1)\hbar^2\psi$, $\hat{J}_Z\psi=M_J\hbar\psi$, and $\hat{J}_z\psi=K\hbar\psi$.)

(2) Some authors, notably Herzberg [38], treat the component quantum numbers Λ, Ω, l and K as taking positive or zero values only, so that each non-zero value of the quantum number labels two wavefunctions with opposite signs for the appropriate angular momentum component. When this is done, lower-case k is often regarded as a signed quantum number, related to K by $K=|k|$. However, in theoretical discussions all component quantum numbers are usually treated as signed, taking both positive and negative values.

(3) There is no uniform convention for denoting the internal vibrational angular momentum; j, π, p and G have all been used. For symmetric top and linear molecules the component of j in the symmetry axis is always denoted by the quantum number l, where l takes values in the range $-v\leqslant l\leqslant +v$ in steps of 2. The corresponding component of angular momentum is actually $l\zeta\hbar$, rather than $l\hbar$, where ζ is a Coriolis coupling constant.

(4) Asymmetric top rotational states are labelled by the value of J (or N if $S\neq 0$), with subscripts K_a, K_c, where the latter correlate with the $K=|k|$ quantum number about the a and c axes in the prolate and oblate symmetric top limits respectively.

Example $J_{Ka,\,Kc}=5_{2,3}$ for a particular rotational level.

Symbols for symmetry operators and labels for symmetry species

(i) *Symmetry operators in space-fixed coordinates* [41]

identity	E
permutation	P
space-fixed inversion	E^*
permutation-inversion	P^* $(=PE^*)$

The permutation operation P permutes the labels of identical nuclei.

Example In the NH_3 molecule, if the hydrogen nuclei are labelled 1, 2 and 3, then $P=(123)$ would symbolize the permutation 1 is replaced by 2, 2 by 3, and 3 by 1.

The inversion operation E^* reverses the sign of all particle coordinates in the space-fixed origin, or in the molecule-fixed centre of mass if translation has been separated. It is also called the parity operator; in field-free space, wavefunctions are either parity + (unchanged) or parity − (change sign) under E^*. The label is used to distinguish between the two nearly degenerate components formed by K-, Λ-, or l-doubling; it is also used in conjunction with the total angular momentum J to define the symmetry labels e and f on parity doublets: see [42].

(ii) *Symmetry operators in molecule fixed coordinates* [38]

identity	E
rotation by $2\pi/n$	C_n
reflection	σ, σ_v, σ_d, σ_h
inversion	i
rotation-reflection	S_n $(=C_n\sigma_h)$

If C_n is the primary axis of symmetry, wavefunctions that are unchanged or change sign under C_n are given species labels A or B respectively, and wavefunctions that are multiplied by $\exp(\pm 2\pi i\ s/n)$ are given the species label E_s. Wavefunctions that are unchanged or change sign under i are labelled g (gerade) or u (ungerade) respectively. Wavefunctions that are unchanged or change sign under σ_h have species labels with a ′ or ″ respectively. For more detailed rules see [37, 38].

Other symbols and conventions in optical spectroscopy

(i) *Term symbols for atomic states*

The electronic states of atoms are labelled by the value of the quantum number L for the state. The value of L is indicated by an upright capital letter: S, P, D, F, G, H, I, and K, . . . , are used for $L = 0$, 1, 2, 3, 4, 5, 6, and 7, . . . , respectively. The corresponding lower-case letters are used for the orbital angular momentum of a single electron. For a many-electron atom, the electron spin multiplicity $(2S+1)$ may be indicated as a left-hand superscript to the letter, and the value of the total angular momentum J as a right-hand subscript. If either L or S is zero only one value of J is possible, and the subscript is then usually suppressed. Finally, the electron configuration of an atom is indicated by giving the occupation of each one-electron orbital as in the examples below.

Examples B: $(1s)^2(2s)^2(2p)^1$, $^2P_{1/2}$
C: $(1s)^2(2s)^2(2p)^2$, $^3P_0$
N: $(1s)^2(2s)^2(2p)^3$, ^4S

(ii) *Term symbols for molecular states*

The electronic states of molecules are labelled by the symmetry species label of the wavefunction in the molecular point group. These should be Latin or Greek upright capital letters. As for atoms, the spin multiplicity $(2S+1)$ may be indicated by a left superscript. For linear molecules the value of Ω $(=\Lambda+\Sigma)$ may be added as a right subscript (analogous to J for atoms). If the value of Ω is not specified, the term symbol is taken to refer to all component states, and a right subscript r or i may be added to indicate that the components are regular (energy increases with Ω) or inverted (energy decreases with Ω) respectively.

The electronic states of molecules are also given empirical single letter labels as follows. The ground electronic state is labelled X, excited states of the same multiplicity are labelled A, B, C, . . . , in ascending order of energy, and excited states of different multiplicity are labelled with lower-case letters a, b, c, . . . In polyatomic molecules (but not diatomic molecules) it is customary to add a tilde (e.g. $\tilde{\mathrm{X}}$) to these empirical labels to prevent possible confusion with the symmetry species label.

Finally the one-electron orbitals are labelled by the corresponding lower-case letters, and the electron configuration is indicated in a manner analogous to that for atoms.

Examples The ground state of CH is $(1\sigma)^2(2\sigma)^2(3\sigma)^2(1\pi)^1$, X $^2\Pi_r$, in which the $^2\Pi_{1/2}$ component lies below the $^2\Pi_{3/2}$ component, as indicated by the subscript r for regular.

The ground state of OH is $(1\sigma)^2(2\sigma)^2(3\sigma)^2(1\pi)^3$, X $^2\Pi_i$, in which the $^2\Pi_{3/2}$ component lies below the $^2\Pi_{1/2}$ component, as indicated by the subscript i for inverted.

The two lowest electronic states of CH_2 are $\ldots(2a_1)^2(1b_2)^2(3a_1)^2$, $\tilde{a}\ ^1A_1$,
$\ldots(2a_1)^2(1b_2)^2(3a_1)^1(1b_1)^1$, $\tilde{X}\ ^3B_1$

The ground state of C_6H_6 (benzene) is $\ldots(a_{2u})^2(e_{1g})^4$, $\tilde{X}\ ^1A_{1g}$.

The vibrational states of molecules are usually indicated by giving the vibrational quantum numbers for each normal mode.

Examples for a bent triatomic molecule,
(0, 0, 0) denotes the ground state,
(1, 0, 0) denotes the ν_1 state, i.e. $v_1 = 1$, and
(1, 2, 0) denotes the $\nu_1 + 2\nu_2$ state, etc.

(iii) *Notation for spectroscopic transitions*
The upper and lower levels of a spectroscopic transition are indicated by a prime ′ and double-prime ″ respectively.

Example $h\nu = E' - E''$

Transitions are generally indicated by giving the excited state label, followed by the ground state label, separated by a dash or an arrow to indicate the direction of the transition (emission to the right, absorption to the left).

Examples B–A indicates a transition between a higher energy state B and a lower energy state A;
B→A indicates emission from B to A;
B←A indicates absorption from A to B.
(0, 2, 1)←(0, 0, 1) labels the $2\nu_2 + \nu_3 - \nu_3$ hot band in a bent triatomic molecule.

A more compact notation [43] may be used to label vibronic (or vibrational) transitions in polyatomic molecules with many normal modes, in which each vibration index r is given a superscript v_r' and a subscript v_r'' indicating the upper and lower state values of the quantum number. When $v_r' = v_r'' = 0$ the corresponding index is suppressed.

Examples 1_0^1 denotes the transition (1, 0, 0) – (0, 0, 0);
$2_0^2\ 3_1^1$ denotes the transition (0, 2, 1) – (0, 0, 1).

For rotational transitions, the value of $\Delta J = J' - J''$ is indicated by a letter labelling the branches of a rotational band: $\Delta J = -2$, -1, 0, 1, and 2 are labelled as the O-branch, P-branch, Q-branch, R-branch, and S-branch respectively. The changes in other quantum numbers (such as K for a symmetric top, or K_a and K_c for an asymmetric top) may be indicated by adding lower-case letters as a left superscript according to the same rule.

Example ^pQ labels a 'p-type Q-branch' in a symmetric top molecule, i.e. $\Delta K = -1$, $\Delta J = 0$.

(iv) *Presentation of spectra*
It is recommended to plot both infra-red and visible/ultraviolet spectra against wavenumber, usually in cm^{-1}, with decreasing wavenumber to the right (note the mnemonic 'red to the right') [9, 20]. (Visible/ultraviolet spectra are also sometimes plotted against wavelength, usually in nm, with

increasing wavelength to the right.) It is recommended to plot Raman spectra with increasing wavenumber shift to the left [10].

It is recommended to plot both electron spin resonance (ESR) spectra and nuclear magnetic resonance (NMR) spectra with increasing magnetic induction (loosely called magnetic field) to the right for fixed frequency, or with increasing frequency to the left for fixed magnetic field [11, 12].

It is recommended to plot photoelectron spectra with increasing ionization energy to the left, i.e. with increasing photoelectron kinetic energy to the right [13].

2.7 ELECTROMAGNETIC RADIATION

The quantities and symbols given here have been selected on the basis of recommendations by IUPAP [3], ISO [4.g], and IUPAC [17, 18, 19].

Name	*Symbol*	*Definition*	*SI unit*	*Notes*
wavelength	λ		m	
speed of light				
in vacuum	c_0		$m\ s^{-1}$	
in a medium	c	$c = c_0/n$	$m\ s^{-1}$	
wavenumber in vacuum	$\tilde{\nu}$	$\tilde{\nu} = \nu/c_0 = 1/n\lambda$	m^{-1}	1
wavenumber (in a medium)	σ	$\sigma = 1/\lambda$	m^{-1}	
frequency	ν	$\nu = c/\lambda$	Hz	
circular frequency, pulsatance	ω	$\omega = 2\pi\nu$	s^{-1}, rad s^{-1}	
refractive index	n	$n = c_0/c$	1	
Planck constant	h		J s	
Planck constant/2π	$\hbar$	$\hbar = h/2\pi$	J s	
radiant energy	Q, W		J	2
radiant energy density	ρ, w	$\rho = Q/V$	$J\ m^{-3}$	2
spectral radiant energy density				2
in terms of frequency	ρ_ν, w_ν	$\rho_\nu = d\rho/d\nu$	$J\ m^{-3}\ Hz^{-1}$	
in terms of wavenumber	$\rho_{\tilde{\nu}}$, $w_{\tilde{\nu}}$	$\rho_{\tilde{\nu}} = d\rho/d\tilde{\nu}$	$J\ m^{-2}$	
in terms of wavelenglth	ρ_λ, w_λ	$\rho_\lambda = d\rho/d\lambda$	$J\ m^{-4}$	
Einstein transition probabilities				3
spontaneous emission	A_{nm}	$dN_n/dt = -A_{nm}N_n$	s^{-1}	
stimulated emission	B_{nm}	$dN_n/dt = -\rho_{\tilde{\nu}}(\tilde{\nu}_{nm}) \times B_{nm}N_n$	$s\ kg^{-1}$	
stimulated absorption	B_{mn}	$dN_n/dt = \rho_{\tilde{\nu}}(\tilde{\nu}_{nm})B_{mn}N_m$	$s\ kg^{-1}$	
radiant power, radiant energy per time	Φ, P	$\Phi = dQ/dt$	W	2
radiant intensity	I	$I = d\Phi/d\Omega$	$W\ sr^{-1}$	2
radiant exitance, (emitted radiant flux)	M	$M = d\Phi/dA_{source}$	$W\ m^{-2}$	2

(1) The unit cm^{-1} is generally used for wavenumber in vacuum.

(2) The symbols for the quantities *radiant energy* through *irradiance* are also used for the corresponding quantities concerning visible radiation, i.e. luminous quantities and photon quantities. Subscripts e for energetic, v for visible, and p for photon may be added whenever confusion between these quantities might otherwise occur. The units used for luminous quantities are derived from the base unit candela (cd), see chapter 3.

Example radiant intensity I_e, SI unit: $W\ sr^{-1}$
luminous intensity I_v, SI unit: cd
photon intensity I_p, SI units: $s^{-1}\ sr^{-1}$

(3) Note that $E_n > E_m$, $E_n - E_m = hc\tilde{\nu}_{nm}$, and $B_{nm} = B_{mn}$, in the definitions. The coefficients B are defined here using energy density $\rho_{\tilde{\nu}}$ in terms of wavenumber; they may alternatively be defined using energy density in terms of frequency ρ_ν, in which case B has SI units $m\ kg^{-1}$.

Name	*Symbol*	*Definition*	*SI unit*	*Notes*
irradiance, (radiant flux received)	E, (I)	$E=\mathrm{d}\Phi/\mathrm{d}A$	$\mathrm{W\,m^{-2}}$	2, 4
emittance	ε	$\varepsilon=M/M_{\mathrm{bb}}$	1	5
Stefan–Boltzmann constant	σ	$M_{\mathrm{bb}}=\sigma T^4$	$\mathrm{W\,m^{-2}\,K^{-4}}$	5
first radiation constant	c_1	$c_1=2\pi hc_0{}^2$	$\mathrm{W\,m^2}$	
second radiation constant	c_2	$c_2=hc_0/k$	$\mathrm{K\,m}$	
transmittance, transmission factor	τ, T	$\tau=\Phi_{\mathrm{tr}}/\Phi_0$	1	6
absorptance, absorption factor	α	$\alpha=\Phi_{\mathrm{abs}}/\Phi_0$	1	6
reflectance, reflection factor	ρ	$\rho=\Phi_{\mathrm{refl}}/\Phi_0$	1	6
(decadic) absorbance	A	$A=-\lg(1-\alpha_{\mathrm{i}})$	1	7
napierian absorbance	B	$B=-\ln(1-\alpha_{\mathrm{i}})$	1	7
absorption coefficient				
(linear) decadic	a, K	$a=A/l$	$\mathrm{m^{-1}}$	8
(linear) napierian	α	$\alpha=B/l$	$\mathrm{m^{-1}}$	8
molar (decadic)	ε	$\varepsilon=a/c=A/cl$	$\mathrm{m^2\,mol^{-1}}$	8, 9
molar napierian	κ	$\kappa=\alpha/c=B/cl$	$\mathrm{m^2\,mol^{-1}}$	8, 10
absorption index	k	$k=\alpha/4\pi\tilde{\nu}$	1	
complex refractive index	$\hat{n}$	$\hat{n}=n+\mathrm{i}k$	1	
molar refraction	R, R_{m}	$R=\dfrac{(n^2-1)}{(n^2+2)}V_{\mathrm{m}}$	$\mathrm{m^3\,mol^{-1}}$	
angle of optical rotation	α		1, rad	11

(4) The irradiance, with SI unit $\mathrm{W\,m^{-2}}$, is often also called the intensity and denoted with the symbol I. This is particularly true in discussions involving collimated beams of light, as in applications of the Lambert–Beer law for spectrometric analysis.
(5) The emittance of a sample is the ratio of the flux emitted by the sample to the flux emitted by a black body at the same temperature; M_{bb} is the latter quantity.
(6) If scattering and luminescence can be neglected, $\tau+\alpha+\rho=1$. In optical spectroscopy internal properties (denoted by a subscript i) are defined to exclude surface effects and effects of the cuvette, so that if scattering and luminescence can be neglected $\tau_{\mathrm{i}}+\alpha_{\mathrm{i}}=1$. This leads to the customary form of the Lambert–Beer law $\Phi_{\mathrm{tr}}/\Phi_0(=I_{\mathrm{tr}}/I_0)=\tau_{\mathrm{i}}=1-\alpha_{\mathrm{i}}=\exp(-\kappa cl)$.
(7) The definitions given here relate the absorbance A or B to the internal absorptance α_{i}; see note (6). However the subscript i on the absorptance α is often omitted.
(8) l is the absorbing path length, and c is the amount (of substance) concentration.
(9) The molar decadic absorption coefficient ε is frequently called the 'extinction coefficient' in published literature. Unfortunately numerical values of the 'extinction coefficient' are often quoted without specifying units; the absence of units usually means that the units used are $\mathrm{mol^{-1}\,dm^3\,cm^{-1}}$. See also [20]. The word 'extinction' should properly be reserved for the sum of the effects of absorption, scattering and luminescence.
(10) κ may be called the molar absorbance cross-section, and κ/N_{A} the cross-section per molecule, denoted σ; this terminology is more common in physics than chemistry.
(11) The optical rotatory power of a solute in solution may be specified by a statement of the type

$$\alpha(589.3\ \mathrm{nm},\ 20\,^{\circ}\mathrm{C},\ \text{sucrose},\ 10\ \mathrm{g\,dm^{-3}}\ \text{in}\ \mathrm{H_2O},\ 10\ \mathrm{cm})=+66.470^{\circ}$$

Note 11 (*continued*)

The same information may also be conveyed by quoting the specific optical rotatory power of the material, usually denoted $[\alpha]_\lambda^\theta$, where

$$[\alpha]_\lambda^\theta = \alpha/\gamma l$$

Here α is the angle of optical rotation, γ the mass concentration, and l the path length. Thus γl is the mass concentration per unit area in the beam path. (Amount concentration c might be more appropriate than mass concentration γ, but in practice it is usually γ that is used.) The specified wavelength λ (frequently the sodium D line) and Celsius temperature θ are written as a subscript and superscript respectively to the specific rotatory power $[\alpha]$. For pure substances $[\alpha]_\lambda^\theta$ is defined similarly by

$$[\alpha]_\lambda^\theta = \alpha/\rho l$$

where ρ is the mass density.

Specific optical rotatory powers are customarily called *specific rotations*. For pure liquids and solutions they are usually quoted as numerical values in units deg dm^{-1} g^{-1} cm^3, where deg is used to symbolize 1° of plane angle. (Sometimes this unit is wrongly described as deg dm^{-1}.) For solids the unit deg mm^{-1} g^{-1} cm^3 is usually used (sometimes wrongly described as: deg mm^{-1}). Neither notation follows the rules of quantity calculus (see chapter 1) because the symbol $[\alpha]$ is usually used to denote the numerical value in specified units rather than the value of the physical quantity itself.

2.8 SOLID STATE

The quantities and their symbols given here have been selected from more extensive lists in IUPAP [3] and ISO [4.n]. See also the *International Tables for Crystallography*, Volume A [44].

Name	*Symbol*	*Definition*	*SI unit*	*Notes*
lattice vector	$\boldsymbol{R}$, $\boldsymbol{R}_0$		m	
fundamental translation vectors for the crystal lattice	$\boldsymbol{a}_1$; $\boldsymbol{a}_2$; $\boldsymbol{a}_3$, $\boldsymbol{a}$; $\boldsymbol{b}$; $\boldsymbol{c}$	$\boldsymbol{R}=n_1\boldsymbol{a}_1+n_2\boldsymbol{a}_2+n_3\boldsymbol{a}_3$	m	1
(circular) reciprocal lattice vector	$\boldsymbol{G}$	$\boldsymbol{G}\cdot\boldsymbol{R}=2\pi m$	m^{-1}	2
(circular) fundamental translation vectors for the reciprocal lattice	$\boldsymbol{b}_1$; $\boldsymbol{b}_2$; $\boldsymbol{b}_3$, $\boldsymbol{a}^*$; $\boldsymbol{b}^*$; $\boldsymbol{c}^*$	$\boldsymbol{a}_i\cdot\boldsymbol{b}_k=2\pi\delta_{ik}$	m^{-1}	3
lattice plane spacing	d		m	
Bragg angle	θ	$n\lambda=2d\sin\theta$	1, rad	
order of reflection	n		1	
order parameters				
short range	σ		1	
long range	s		1	
Burgers vector	$\boldsymbol{b}$		m	
particle position vector	$\boldsymbol{r}$, $\boldsymbol{R}_j$		m	4
equilibrium position vector of an ion	$\boldsymbol{R}_0$		m	
displacement vector of an ion	$\boldsymbol{u}$	$\boldsymbol{u}=\boldsymbol{R}-\boldsymbol{R}_0$	m	
Debye–Waller factor	B, D		1	
Debye circular wavenumber	q_{D}		m^{-1}	
Debye circular frequency	ω_{D}		s^{-1}	
Grüneisen parameter	γ, Γ	$\gamma=\alpha V/\kappa C_V$	1	5
Madelung constant	α, $\mathcal{M}$	$E_{\mathrm{coul}}=\dfrac{\alpha N_{\mathrm{A}} z_+ z_- e^2}{4\pi\varepsilon_0 R_0}$	1	
density of states	N_E	$N_E=\mathrm{d}N(E)/\mathrm{d}E$	$\mathrm{J}^{-1}\,\mathrm{m}^{-3}$	6
(spectral) density of vibrational modes	N_ω, g	$N_\omega=\mathrm{d}N(\omega)/\mathrm{d}\omega$	$\mathrm{s\,m}^{-3}$	7
resistivity tensor	ρ_{ik}	$\boldsymbol{E}=\boldsymbol{\rho}\cdot\boldsymbol{j}$	$\Omega\,\mathrm{m}$	

(1) n_1, n_2 and n_3 are integers. a, b, c are also called the lattice constants.
(2) m is an integer.
(3) Reciprocal lattice vectors are sometimes defined by $\boldsymbol{a}_i\cdot\boldsymbol{b}_k=\delta_{ik}$.
(4) To distinguish between electron and ion position vectors, lower-case and capital letters are used respectively. The subscript j relates to particle j.
(5) α is the cubic expansion coefficient, V the volume, κ the isothermal compressibility, and C_V the heat capacity at constant volume.
(6) $N(E)$ is the total number of states of electronic energy less than E, divided by the volume.
(7) $N(\omega)$ is the total number of vibrational modes with circular frequency less than ω, divided by the volume.

Name	Symbol	Definition	SI unit	Notes
conductivity tensor	σ_{ik}	$\sigma = \rho^{-1}$	$S\,m^{-1}$	
thermal conductivity tensor	λ_{ik}	$\boldsymbol{J}_q = -\boldsymbol{\lambda} \cdot \mathrm{grad}\, T$	$W\,m^{-1}\,K^{-1}$	
residual resistivity	ρ_R		$\Omega\,m$	
relaxation time	τ	$\tau = l/v_F$	s	8
Lorenz coefficient	L	$L = \lambda/\sigma T$	$V^2\,K^{-2}$	
Hall coefficient	A_H, R_H	$\boldsymbol{E} = \boldsymbol{\rho} \cdot \boldsymbol{j} + R_H(\boldsymbol{B} \times \boldsymbol{j})$	$m^3\,C^{-1}$	
thermoelectric force	E		V	9
Peltier coefficient	Π		V	9
Thomson coefficient	μ, (τ)		$V\,K^{-1}$	
work function	Φ	$\Phi = E_\infty - E_F$	J	10
number density, number concentration	n, (p)		m^{-3}	11
gap energy	E_g		J	
donor ionization energy	E_d		J	
acceptor ionization energy	E_a		J	
Fermi energy	E_F, ε_F		J	
circular wave vector, propagation vector	$\boldsymbol{k}$, $\boldsymbol{q}$	$k = 2\pi/\lambda$	m^{-1}	12
Bloch function	$u_k(\boldsymbol{r})$	$\psi(\boldsymbol{r}) = u_k(\boldsymbol{r}) \exp(\mathrm{i}\boldsymbol{k} \cdot \boldsymbol{r})$	$m^{-3/2}$	13
charge density of electrons	ρ	$\rho(\boldsymbol{r}) = -e\psi^*(\boldsymbol{r})\psi(\boldsymbol{r})$	$C\,m^{-3}$	13, 14
effective mass	m^*		kg	15
mobility	μ	$\mu = v_{\mathrm{drift}}/E$	$m^2\,V^{-1}\,s^{-1}$	15
mobility ratio	b	$b = \mu_n/\mu_p$	1	
diffusion coefficient	D	$\mathrm{d}N/\mathrm{d}t = -DA(\mathrm{d}n/\mathrm{d}x)$	$m^2\,s^{-1}$	15
diffusion length	L	$L = \sqrt{D\tau}$	m	15, 16
characteristic (Weiss) temperature	θ, θ_W		K	
Curie temperature	T_C		K	
Néel temperature	T_N		K	

(8) The definition applies to electrons in metals; l is the mean free path, and v_F is the electron velocity on the Fermi sphere.
(9) The substances to which the symbol applies are denoted by subscripts.
(10) E_∞ is the electron energy at rest at infinite distance.
(11) Specific number densities are denoted by subscripts: for electrons n_n, n_-, (n); for holes n_p, n_+, p; for donors n_d; for acceptors n_a; for the intrinsic number density n_i ($n_i^2 = n_+ n_-$).
(12) $\boldsymbol{k}$ is used for particles, $\boldsymbol{q}$ for phonons.
(13) $\psi(\boldsymbol{r})$ is a one-electron wavefunction.
(14) The total charge density is obtained by summing over all electrons.
(15) Subscripts n and p or − and + may be used to denote electrons and holes respectively.
(16) D is the diffusion coefficient and τ the lifetime.

Symbols for planes and directions in crystals

Miller indices of a crystal face, or of a single net plane	(h, k, l) or (h_1, h_2, h_3)
indices of the Bragg reflection from the set of parallel net planes (h, k, l)	h, k, l or h_1, h_2, h_3
indices of a set of all symmetrically equivalent crystal faces, or net planes	$\{h, k, l\}$ or $\{h_1, h_2, h_3\}$
indices of a lattice direction (zone axis)	$[u, v, w]$
indices of a set of symmetrically equivalent lattice directions	$\langle u, v, w\rangle$

In each of these cases, when the letter symbol is replaced by numbers it is customary to omit the commas. For a single plane or crystal face, or a specific direction, a negative number is indicated by a bar over the number.

Example $(\bar{1}10)$ denotes the set of parallel planes $h=-1$, $k=+1$, $l=0$.

2.9 STATISTICAL THERMODYNAMICS

The names and symbols given here are in agreement with those recommended by IUPAP [3] and by ISO [4.i].

Name	*Symbol*	*Definition*	*SI units*	*Notes*
number of entities	N		1	
number density of entities, number concentration	n, C	$n = N/V$	$\mathrm{m^{-3}}$	
Avogadro constant	L, N_A		$\mathrm{mol^{-1}}$	
Boltzmann constant	k, k_B		$\mathrm{J\,K^{-1}}$	
gas constant (molar)	R	$R = Lk$	$\mathrm{J\,K^{-1}\,mol^{-1}}$	
molecular position vector	$\boldsymbol{r}\,(x, y, z)$		m	
molecular velocity vector	$\boldsymbol{c}(c_x, c_y, c_z)$, $\boldsymbol{u}(u_x, u_y, u_z)$	$\boldsymbol{c} = \mathrm{d}\boldsymbol{r}/\mathrm{d}t$	$\mathrm{m\,s^{-1}}$	
molecular momentum vector	$\boldsymbol{p}(p_x, p_y, p_z)$	$\boldsymbol{p} = m\boldsymbol{c}$	$\mathrm{kg\,m\,s^{-1}}$	
velocity distribution function (Maxwell)	$f(c_x)$	$f(c_x) = (m/2\pi kT)^{\frac{1}{2}} \times \exp(-mc_x{}^2/2kT)$	$\mathrm{m^{-1}\,s}$	
speed distribution function (Maxwell–Boltzmann)	$F(c)$	$F(c) = (m/2\pi kT)^{3/2} \times 4\pi c^2 \exp(-mc^2/2kT)$	$\mathrm{m^{-1}\,s}$	
average speed	$\bar{c}$, $\bar{u}$, $\langle c\rangle$, $\langle u\rangle$	$\bar{c} = \int cF(c)\mathrm{d}c$	$\mathrm{m\,s^{-1}}$	
generalized coordinate	q		(m)	1
generalized momentum	p	$p = \partial L/\partial \dot{q}$	$(\mathrm{kg\,m\,s^{-1}})$	1
volume in phase space	Ω	$\Omega = (1/h)\int p\,\mathrm{d}q$	1	
probability	P		1	
statistical weight, degeneracy	g, d, W, ω, β		1	2
density of states	$\rho(E)$	$\rho(E) = \mathrm{d}N/\mathrm{d}E$	$\mathrm{J^{-1}}$	
partition function, sum over states,				
for a single molecule	q, z	$q = \sum_i g_i \exp(-\varepsilon_i/kT)$	1	3
for a canonical ensemble (system, or assembly)	Q, Z		1	
microcanonical ensemble	Ω		1	
grand (canonical ensemble)	Ξ		1	

(1) If q is a length then p is a momentum as indicated by the units in parentheses. In the definition of p, L denotes the Lagrangian.
(2) β is usually used for a spin statistical weight.
(3) ε_i denotes the energy of the ith molecular level.

Name	*Symbol*	*Definition*	*SI unit*	*Notes*
symmetry number	σ, s		1	
reciprocal temperature parameter	β	$\beta = 1/kT$	J^{-1}	
characteristic temperature	Θ		K	4

(4) Particular characteristic temperatures are denoted with subscripts, e.g. rotational $\Theta_r = hc\tilde{B}/k$, vibrational $\Theta_v = hc\tilde{\nu}/k$, Debye $\Theta_D = hc\tilde{\nu}_D/k$, Einstein $\Theta_E = hc\tilde{\nu}_E/k$.

2.10 GENERAL CHEMISTRY

The symbols given by IUPAP [3] and by ISO [4.e,i] are in agreement with the recommendations given here.

Name	*Symbol*	*Definition*	*SI unit*	*Notes*
number of entities (e.g. molecules, atoms, ions, formula units)	N		1	
amount (of substance)	n	$n_B = N_B/L$	mol	1, 2
Avogadro constant	L, N_A		mol^{-1}	
mass of atom, atomic mass	m_a, m		kg	
mass of entity (molecule, or formula unit)	m_f, m		kg	
atomic mass constant	m_u	$m_u = m_a(^{12}C)/12$	kg	3
molar mass	M	$M_B = m/n_B$	$kg\ mol^{-1}$	2, 4
relative molecular mass (relative molar mass, molecular weight)	M_r	$M_{r,B} = m_B/m_u$	1	2, 5
molar volume	V_m	$V_{m,B} = V/n_B$	$m^3\ mol^{-1}$	2, 4
mass fraction	w	$w_B = m_B/\Sigma m_i$	1	2
volume fraction	ϕ	$\phi_B = V_B/\Sigma V_i$	1	2, 6
mole fraction, amount fraction, number fraction	x, y	$x_B = n_B/\Sigma n_i$	1	2, 7
(total) pressure	p, P		Pa	
partial pressure	p_B	$p_B = y_B p$	Pa	8
mass concentration (mass density)	γ, ρ	$\gamma_B = m_B/V$	$kg\ m^{-3}$	2, 9, 10

(1) The words 'of substance' may be replaced by the specification of the entity.

Example When the amount of O_2 is equal to 3 moles, $n(O_2) = 3$ mol, then the amount of $\frac{1}{2}O_2$ is equal to 6 moles, $n(\frac{1}{2}O_2) = 6$ mol. Thus $n(\frac{1}{2}O_2) = 2n(O_2)$. See also p.41 below.

(2) The definition applies to entities B which should always be indicated by a subscript or in parentheses, e.g. n_B or $n(B)$.
(3) m_u is equal to the unified atomic mass unit, with symbol u, i.e. $m_u = 1$ u (see section 3.7). In biochemistry this unit is called the dalton, with symbol Da, although the name and symbol have not been approved by CGPM.
(4) The definition applies to pure substance, where m is the total mass and V is the total volume. However, corresponding quantities may also be defined for a mixture as m/n and V/n, where $n = \sum_i n_i$. These quantities are called the mean molar mass and the mean molar volume respectively.
(5) For molecules M_r is the relative molecular mass or molecular weight; for atoms M_r is the relative atomic mass or atomic weight and the symbol A_r may be used. M_r may also be called the relative molar mass, $M_{r,B} = M_B/(g\ mol^{-1})$.
(6) V_B and V_i are the volumes of appropriate components prior to mixing.
(7) For condensed phases x is used, and for gaseous mixtures y may be used.
(8) The symbol and the definition apply to entities B, which should be specified.
(9) V is the volume of the mixture.
(10) In polymer science the symbol c is often used for mass concentration.

Name	*Symbol*	*Definition*	*SI unit*	*Notes*
number concentration, number density of entities	C, n	$C_B = N_B/V$	m^{-3}	2, 9, 11
amount concentration, concentration	c	$c_B = n_B/V$	$mol\ m^{-3}$	2, 9, 12
solubility	s	$s_B = c_B$ (saturated solution)	$mol\ m^{-3}$	2
molality (of a solute)	m, (b)	$m_B = n_B/m_A$	$mol\ kg^{-1}$	2, 13
surface concentration	Γ	$\Gamma_B = n_B/A$	$mol\ m^{-2}$	2
stoichiometric number	ν		1	14
extent of reaction, advancement	ξ	$\Delta\xi = \Delta n_B/\nu_B$	mol	2, 15
degree of dissociation	α		1	

(11) The term number concentration and symbol C is preferred for mixtures.
(12) 'Amount concentration' is an abbreviation for 'amount-of-substance concentration'. When there is no risk of confusion the word 'concentration' may be used alone. The symbol [B] is often used for amount concentration of entities B. This quantity is also sometimes called molarity. A solution of, for example, 1 mol dm^{-3} is often called a 1 molar solution, denoted 1 M solution. However, M should not be treated as a symbol for the unit mol dm^{-3}, in the sense that it should not be used with SI prefixes or in conjunction with other units.
(13) In the definition m_B denotes the molality of solute B, and m_A denotes the mass of solvent A; thus the same symbol m is used with two different meanings. This confusion of notation may be avoided by using the symbol b for molality, but this is seldom done.

A solution of molality 1 mol/kg is occasionally called a 1 molal solution, denoted 1 m solution; however, the symbol m should not be treated as a symbol for the unit mol kg^{-1}.
(14) The stoichiometric number is defined through the reaction equation. It is negative for reactants and positive for products. The values of the stoichiometric numbers depend on how the reaction equation is written.

Example $(1/2)N_2 + (3/2)H_2 = NH_3$: $\nu(N_2) = -1/2$,
$\nu(H_2) = -3/2$,
$\nu(NH_3) = +1$.

(15) The extent of reaction also depends on how the reaction equation is written, but it is independent of which entity in the reaction equation is used in the definition.

Example For the reaction in footnote (14), when $\Delta\xi = 2$ mol, $\Delta n(N_2) = -1$ mol, $\Delta n(H_2) = -3$ mol, and $\Delta n(NH_3) = +2$ mol.

This quantity was originally introduced as *degré d'avancement* by de Donder.

Other symbols and conventions in chemistry

(i) *Symbols for particles and nuclear reactions*

neutron	n	helion	h
proton	p	alpha particle	α
deuteron	d	electron	e
triton	t	photon	γ
positive muon	μ^+	negative muon	μ^-

The electric charge of particles may be indicated by adding the superscript +, −, or 0; e.g. p^+, n^0, e^-, etc. If the symbols p and e are used without a charge, they refer to the positive proton and negative electron respectively.

The meaning of the symbolic expression indicating a nuclear reaction should be as follows:

$$\begin{matrix}\text{initial} \\ \text{nuclide}\end{matrix}\left(\begin{matrix}\text{incoming particles} \\ \text{or quanta}\end{matrix}, \begin{matrix}\text{outgoing particles} \\ \text{or quanta}\end{matrix}\right)\begin{matrix}\text{final} \\ \text{nuclide}\end{matrix}$$

Examples $^{14}\mathrm{N}(\alpha,\mathrm{p})^{17}\mathrm{O}$, $^{59}\mathrm{Co}(\mathrm{n},\gamma)^{60}\mathrm{Co}$,
$^{23}\mathrm{Na}(\gamma,3\mathrm{n})^{20}\mathrm{Na}$, $^{31}\mathrm{P}(\gamma,\mathrm{pn})^{29}\mathrm{Si}$

(ii) *Chemical symbols for the elements*

The chemical symbols of elements are (in most cases) derived from their Latin names and consist of one or two letters which should always be printed in roman (upright) type. Only for elements of atomic number greater than 103, the systematic symbols consist of three letters (see footnote U to table 6.2). A complete list is given in table 6.2. The symbol is not followed by a full stop except at the end of a sentence.

Examples I, U, Pa, C

The symbols can have different meanings:

(a) They can denote an atom of the element. For example, Cl can denote a chlorine atom having 17 protons and 18 or 20 neutrons (giving a mass number of 35 or 37), the difference being ignored. Its mass is on average 35.4527 u in terrestrial samples.

(b) The symbol may, as a kind of shorthand, denote a sample of the element. For example, Fe can denote a sample of iron, and He a sample of helium gas.

The term *nuclide* implies an atom of specified atomic number (proton number) and mass number (nucleon number). Nuclides having the same atomic number but different mass numbers are called isotopic nuclides or *isotopes*. Nuclides having the same mass number but different atomic numbers are called isobaric nuclides or *isobars*.

A nuclide may be specified by attaching the mass number as a left superscript to the symbol for the element.The atomic number may also be attached as a left subscript, if desired, although this is rarely done. If no left superscript is attached, the symbol is read as including all isotopes in natural abundance.

Examples $^{14}\mathrm{N}$, $^{12}\mathrm{C}$, $^{13}\mathrm{C}$, $^{16}_{8}\mathrm{O}$, $n(\mathrm{Cl})=n(^{35}\mathrm{Cl})+n(^{37}\mathrm{Cl})$

The ionic charge number is denoted by a right superscript, the sign being given after its absolute value (which may be omitted when equal to 1).

Examples Na^{+}	a sodium positive ion (cation)
$^{79}\mathrm{Br}^{-}$	a bromine-79 negative ion (anion, bromide ion)
$3\ \mathrm{Al}^{3+}$	three aluminium triply positive ions
S^{2-}	a sulfur doubly negative ion (sulfide ion)

The right superscript position is also used to convey other information: Excited electronic states may be denoted by an asterisk,

Examples H^{*}, Cl^{*}

Oxidation numbers are denoted by positive or negative roman numerals or by zero (see also (iv) below),

Examples $\mathrm{Mn}^{\mathrm{VII}}$, $\mathrm{O}^{-\mathrm{II}}$, Ni^{0}

The positions and meanings of indices around the symbol of the element are summarized as follows.

left superscript	mass number
left subscript	atomic number
right superscript	charge number, oxidation number, excitation symbol
right subscript	number of atoms per entity (see (iii) below)

(iii) *Chemical formulae*
Chemical formulae denote entities composed of more than one atom (molecules, complex ions, groups of atoms, etc.).

Examples N_2, P_4, C_6H_6, $CaSO_4$, $PtCl_4{}^{2-}$, $Fe_{0.91}S$

They may also be used as a shorthand to denote a sample of the corresponding chemical substance.

Examples	CH_3OH	methanol
	$\rho(H_2SO_4)$	mass density of sulfuric acid

The number of atoms in an entity is indicated by a right subscript (the numeral 1 being omitted). Groups of atoms may also be enclosed in parentheses. Entities may be specified by giving the corresponding formula, often multiplied by a factor. Charge numbers of complex ions, and excitation symbols, are added as right superscripts to the whole formula. The free radical nature of some entities may be stressed by adding a dot to the symbol.

Examples	H_2O	one water molecule, water
	$1/2\ O_2$	half an oxygen molecule
	$Zn_3(PO_4)_2$	one zinc phosphate formula unit, zinc phosphate
	$2\ MgSO_4$	two formula units of magnesium sulfate
	$1/5\ KMnO_4$	one-fifth of a potassium permanganate formula unit
	$1/2\ SO_4{}^{2-}$	half a sulfate ion
	$CH_3\cdot$	one methyl radical
	$CH_3\dot{C}HCH_3$	isopropyl radical
	$NO_2{}^*$	electronically excited nitrogen dioxide molecule

In the above examples, $\frac{1}{2}O_2$, $1/5\ KMnO_4$ and $\frac{1}{2}SO_4{}^{2-}$ are artificial in the sense that such fractions of a molecule cannot exist. However, it may often be convenient to specify entities this way when calculating amounts of substance; see (v) below.

Specific electronic states of entities (atoms, molecules, ions) can be denoted by giving the electronic term symbol (see section 2.6) in parentheses. Vibrational and rotational states can be specified by giving the corresponding quantum numbers.

Examples	$Hg(^3P_1)$	a mercury atom in the triplet-P-one state
	$HF(v=2,\ J=6)$	a hydrogen fluoride molecule in the vibrational state $v=2$ and the rotational state $J=6$
	$H_2O^+(^2A_1)$	a water molecule ion in the doublet-A-one state

Chemical formulae may be written in different ways according to the information that they convey, as follows:

Formula	*Information conveyed*	*Example for lactic acid*
empirical	stoichiometric proportion only	CH_2O
molecular	in accord with molecular mass	$C_3H_6O_3$
structural	structural relationship of atoms	$CH_3CHOHCOOH$
displayed	projection of atoms and bonds	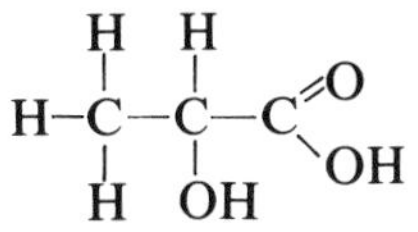
stereochemical	stereochemical relationship	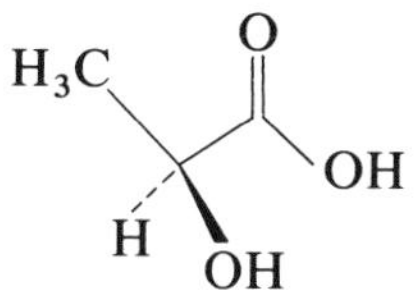

Further conventions for writing chemical formulae are described in [21, 22].

(iv) *Equations for chemical reactions*

Symbols connecting the reactants and products in a chemical reaction equation have the following meanings:

$H_2 + Br_2 \rightarrow 2HBr$	forward reaction
$H_2 + Br_2 \leftrightarrows 2HBr$	reaction, both directions
$H_2 + Br_2 \rightleftharpoons 2HBr$	equilibrium
$H_2 + Br_2 = 2HBr$	stoichiometric relation

If a reaction is considered elementary, for example $H + Br_2 \rightarrow HBr + Br$, it should be made clear in the text.

Redox equations are often written so that the absolute value of the stoichiometric coefficient for the electrons transferred (which are normally omitted from the overall equation) is equal to one.

Example $1/5\, KMn^{VII}O_4 + 8/5\, HCl = 1/5\, Mn^{II}Cl_2 + 1/2\, Cl_2 + 1/5\, KCl + 4/5\, H_2O$

Similarly a reaction in an electrochemical cell may be written so that the charge number of the cell reaction is equal to one:

Example $1/3\, In^0(s) + 1/2\, Hg^{I}{}_2SO_4(s) = 1/6\, In^{III}{}_2(SO_4)_3(aq) + Hg^0(l)$

(the symbols in parentheses denote the state, see (vi) below).

(v) *Amount of substance and the specification of entities*

The quantity 'amount of substance' has been used by chemists for a long time without a proper name. It was simply referred to as the 'number of moles'. This practice should be abandoned, because it is wrong to confuse the name of a physical quantity with the name of a unit (in a similar way it would be wrong to use 'number of metres' as a synonym for 'length'). The amount of substance is proportional to the number of specified elementary entities of that substance; the proportionality factor is the same for all substances and is the reciprocal of the Avogadro constant. The elementary entities may be chosen as convenient, not necessarily as physically real individual particles. Since the amount of substance and all physical quantities derived from it depend on this choice it is essential to specify the entities to avoid ambiguities.

Examples	n_{Cl}, $n(Cl)$	amount of Cl, amount of chlorine atoms
	$n(Cl_2)$	amount of Cl_2, amount of chlorine molecules
	$n(H_2SO_4)$	amount of (entities) H_2SO_4
	$n(1/5\,KMnO_4)$	amount of (entities) $1/5\,KMnO_4$
	$M(P_4)$	molar mass of (tetraphosphorus) P_4
	c_{HCl}, $c(HCl)$, $[HCl]$	amount concentration of HCl
	$\Lambda(MgSO_4)$	molar conductivity of (magnesium sulfate entities) $MgSO_4$
	$\Lambda(\frac{1}{2}MgSO_4)$	molar conductivity of (entities) $\frac{1}{2}MgSO_4$

$n(1/5\,KMnO_4) = 5n\,(KMnO_4)$
$\lambda(\frac{1}{2}Mg^{2+}) = \frac{1}{2}\lambda(Mg^{2+})$
$[\frac{1}{2}H_2SO_4] = 2[H_2SO_4]$
(see also examples in section 3.2, p.64.)

Note that 'amount of sulfur' is an ambiguous statement, because it might imply $n(S)$, $n(S_8)$, or $n(S_2)$, etc. In some cases analogous statements are less ambiguous. Thus for compounds the implied entity is usually the molecule or the common formula entity, and for solid metals it is the atom.

Examples '2 moles of water' implies $n(H_2O) = 2$ mol; '0.5 moles of sodium chloride' implies $n(NaCl) = 0.5$ mol; '3 millimoles of iron' implies $n(Fe) = 3$ mmol, but such statements should be avoided whenever there might be ambiguity.

(vi) *States of aggregation*

The following one-, two- or three-letter symbols are used to represent the states of aggregation of chemical species [1.j]. The letters are appended to the formula symbol in parentheses, and should be printed in roman (upright) type without a full stop (period).

g	gas or vapour	vit	vitreous substance
l	liquid	a, ads	species adsorbed on a substrate
s	solid	mon	monomeric form
cd	condensed phase (i.e. solid or liquid)	pol	polymeric form
		sln	solution
fl	fluid phase (i.e. gas or liquid)	aq	aqueous solution
		aq, ∞	aqueous solution at infinite dilution
cr	crystalline		
lc	liquid crystal	am	amorphous solid

Examples	HCl(g)	hydrogen chloride in the gaseous state
	C_V(fl)	heat capacity of a fluid at constant volume
	V_m(lc)	molar volume of a liquid crystal
	U(cr)	internal energy of a crystalline solid
	MnO_2(am)	manganese dioxide as an amorphous solid
	MnO_2(cr, I)	manganese dioxide as crystal form I
	NaOH(aq)	aqueous solution of sodium hydroxide
	NaOH(aq, ∞)	. . . as above, at infinite dilution
	$\Delta_f H^\circ(H_2O, l)$	standard enthalpy of formation of liquid water

The symbols g, l, to denote gas phase, liquid phase, etc., are also sometimes used as a right superscript, and the Greek letter symbols α, β, may be similarly used to denote phase α, phase β, etc., in a general notation.

Examples $V_m^{\,l}$, $V_m^{\,s}$ molar volume of the liquid phase, . . . of the solid phase
$S_m^{\,\alpha}$, $S_m^{\,\beta}$ molar entropy of phase α, . . . of phase β

2.11 CHEMICAL THERMODYNAMICS

The names and symbols of the more generally used quantities given here are also recommended by IUPAP [3] and by ISO [4.e,i]. Additional information can be found in [1.d,j].

Name	*Symbol*	*Definition*	*SI unit*	*Notes*
heat	q, Q		J	1
work	w, W		J	1
internal energy	U	$\Delta U = q + w$	J	1
enthalpy	H	$H = U + pV$	J	
thermodynamic temperature	T		K	
Celsius temperature	θ, t	$\theta/°\mathrm{C} = T/\mathrm{K} - 273.15$	°C	2
entropy	S	$\mathrm{d}S \geqslant \mathrm{d}q/T$	$\mathrm{J\,K^{-1}}$	
Helmholtz energy, (Helmholtz function)	A	$A = U - TS$	J	3
Gibbs energy, (Gibbs function)	G	$G = H - TS$	J	
Massieu function	J	$J = -A/T$	$\mathrm{J\,K^{-1}}$	
Planck function	Y	$Y = -G/T$	$\mathrm{J\,K^{-1}}$	
surface tension	γ, σ	$\gamma = (\partial G/\partial A_\mathrm{s})_{T,p}$	$\mathrm{J\,m^{-2}}$, $\mathrm{N\,m^{-1}}$	
molar quantity X	X_m	$X_\mathrm{m} = X/n$	(varies)	4, 5
specific quantity X	x	$x = X/m$	(varies)	4, 5
pressure coefficient	β	$\beta = (\partial p/\partial T)_V$	$\mathrm{Pa\,K^{-1}}$	
relative pressure coefficient	α_p	$\alpha_p = (1/p)(\partial p/\partial T)_V$	$\mathrm{K^{-1}}$	
compressibility,				
isothermal	κ_T	$\kappa_T = -(1/V)(\partial V/\partial p)_T$	$\mathrm{Pa^{-1}}$	
isentropic	κ_S	$\kappa_S = -(1/V)(\partial V/\partial p)_S$	$\mathrm{Pa^{-1}}$	
linear expansion coefficient	α_l	$\alpha_l = (1/l)(\partial l/\partial T)$	$\mathrm{K^{-1}}$	
cubic expansion coefficient	α, α_V, γ	$\alpha = (1/V)(\partial V/\partial T)_p$	$\mathrm{K^{-1}}$	6
heat capacity,				
at constant pressure	C_p	$C_p = (\partial H/\partial T)_p$	$\mathrm{J\,K^{-1}}$	
at constant volume	C_V	$C_V = (\partial U/\partial T)_V$	$\mathrm{J\,K^{-1}}$	
ratio of heat capacities	γ, (κ)	$\gamma = C_p/C_V$	1	
Joule–Thomson coefficient	μ, μ_JT	$\mu = (\partial T/\partial p)_H$	$\mathrm{K\,Pa^{-1}}$	

(1) Both $q > 0$ and $w > 0$ indicate an increase in the energy of the system; $\Delta U = q + w$.
(2) This quantity is sometimes misnamed 'centigrade temperature'.
(3) It is sometimes convenient to use the symbol F for Helmholtz energy in the context of surface chemistry, to avoid confusion with A for area.
(4) The definition applies to pure substance. However, the concept of molar and specific quantities may also be applied to mixtures.
(5) X is an extensive quantity. The unit depends on the quantity. In the case of molar quantities the entities should be specified.

Example molar volume of B, $V_\mathrm{m}(\mathrm{B}) = V/n_\mathrm{B}$

(6) This quantity is also called the coefficient of thermal expansion, or the expansivity coefficient.

Name	*Symbol*	*Definition*	*SI unit*	*Notes*
second virial coefficient	B	$pV_m = RT(1 + B/V_m + \cdots)$	$m^3 mol^{-1}$	
compression factor (compressibility factor)	Z	$Z = pV_m/RT$	1	
partial molar quantity X	X_B, (X'_B)	$X_B = (\partial X/\partial n_B)_{T, p, n_{j \neq B}}$	(varies)	7
chemical potential (partial molar Gibbs energy)	μ	$\mu_B = (\partial G/\partial n_B)_{T, p, n_{j \neq B}}$	$J mol^{-1}$	8
absolute activity	λ	$\lambda_B = \exp(\mu_B/RT)$	1	8
standard chemical potential	$\mu^{\ominus}$, μ^{o}		$J mol^{-1}$	9
standard partial molar enthalpy	$H_B{}^{\ominus}$	$H_B{}^{\ominus} = \mu_B{}^{\ominus} + TS_B{}^{\ominus}$	$J mol^{-1}$	8, 9
standard partial molar entropy	$S_B{}^{\ominus}$	$S_B{}^{\ominus} = -(\partial \mu_B{}^{\ominus}/\partial T)_p$	$J mol^{-1} K^{-1}$	8, 9
standard reaction Gibbs energy (function)	$\Delta_r G^{\ominus}$	$\Delta_r G^{\ominus} = \sum_B \nu_B \mu_B{}^{\ominus}$	$J mol^{-1}$	9, 10, 11
affinity of reaction	A, $(\mathscr{A})$	$A = -(\partial G/\partial \xi)_{p, T} = -\sum_B \nu_B \mu_B$	$J mol^{-1}$	11
standard reaction enthalpy	$\Delta_r H^{\ominus}$	$\Delta_r H^{\ominus} = \sum_B \nu_B H_B{}^{\ominus}$	$J mol^{-1}$	9, 10, 11
standard reaction entropy	$\Delta_r S^{\ominus}$	$\Delta_r S^{\ominus} = \sum_B \nu_B S_B{}^{\ominus}$	$J mol^{-1} K^{-1}$	9, 10, 11
equilibrium constant	$K^{\ominus}$, K	$K^{\ominus} = \exp(-\Delta_r G^{\ominus}/RT)$	1	9, 11, 12
equilibrium constant, pressure basis	K_p	$K_p = \prod_B p_B{}^{\nu_B}$	$Pa^{\Sigma\nu}$	11, 13

(7) The symbol applies to entities B which should be specified. The prime may be used to distinguish partial molar X from X when necessary.

Example The partial molar volume of Na_2SO_4 in aqueous solution may be denoted $V'(Na_2SO_4, aq)$, in order to distinguish it from the volume of the solution $V(Na_2SO_4, aq)$.

(8) The definition applies to entities B which should be specified.
(9) The symbol $^{\ominus}$ or ^{o} is used to indicate standard. They are equally acceptable. Definitions of standard states are discussed below (p.47). Whenever a standard chemical potential $\mu^{\ominus}$ or a standard equilibrium constant $K^{\ominus}$ or other standard quantity is used, the standard state must be specified.
(10) The symbol r indicates reaction in general. In particular cases r can be replaced by another appropriate subscript, e.g. $\Delta_f H^{\ominus}$ denotes the standard molar enthalpy of formation; see p.45 below for a list of subscripts.
(11) The reaction must be specified for which this quantity applies.
(12) This quantity is dimensionless and its value depends on the choice of standard state, which must be specified. ISO [4.i] recommend the symbol $K^{\ominus}$ and the name 'standard equilibrium constant'. The IUPAC Thermodynamics Commission [1.j] recommend the symbol K and the name 'thermodynamic equilibrium constant'.
(13) These quantities are not in general dimensionless. One can define in an analogous way an equilibrium constant in terms of fugacity K_f, etc. At low pressures K_p is approximately related to $K^{\ominus}$ by the equation $K^{\ominus} \approx K_p/(p^{\ominus})^{\Sigma\nu}$, and similarly in dilute solutions K_c is approximately related to $K^{\ominus}$ by $K^{\ominus} \approx K_c/(c^{\ominus})^{\Sigma\nu}$; however, the exact relations involve fugacity coefficients or activity coefficients [1.j].

The equilibrium constant of dissolution of an electrolyte (describing the equilibrium between excess solid phase and solvated ions) is often called a solubility product, denoted K_{sol} or K_s (or $K_{sol}{}^{\ominus}$ or $K_s{}^{\ominus}$ as appropriate). In a similar way the equilibrium constant for an acid dissociation is often written K_a, for base hydrolysis K_b, and for water dissociation K_w.

Name	Symbol	Definition	SI Unit	Notes
equilibrium constant (*cont.*)				
concentration basis	K_c	$K_c=\prod_B c_B^{\nu_B}$	$(\text{mol m}^{-3})^{\Sigma\nu}$	11, 13
molality basis	K_m	$K_m=\prod_B m_B^{\nu_B}$	$(\text{mol kg}^{-1})^{\Sigma\nu}$	11, 13
fugacity	$f, \tilde{p}$	$f_B=\lambda_B \lim_{p\to 0} (p_B/\lambda_B)_T$	Pa	8
fugacity coefficient	ϕ	$\phi_B=f_B/p_B$	1	
activity and activity coefficient referenced to Raoult's law,				
(relative) activity	a	$a_B=\exp\left[\frac{\mu_B-\mu_B{}^*}{RT}\right]$	1	8, 14, 15
activity coefficient	f	$f_B=a_B/x_B$	1	8
activities and activity coefficients referenced to Henry's law, (relative) activity,				
molality basis	a_m	$a_{m,B}=\exp\left[\frac{\mu_B-\mu_B{}^\circ}{RT}\right]$	1	8, 15, 16
concentration basis	a_c	$a_{c,B}=\exp\left[\frac{\mu_B-\mu_B{}^\circ}{RT}\right]$	1	8, 15, 16
mole fraction basis	a_x	$a_{x,B}=\exp\left[\frac{\mu_B-\mu_B{}^\circ}{RT}\right]$	1	8, 15, 16
activity coefficient,				
molality basis	γ_m	$a_{m,B}=\gamma_{m,B}m_B/m^\circ$	1	8
concentration basis	γ_c	$a_{c,B}=\gamma_{c,B}c_B/c^\circ$	1	8
mole fraction basis	γ_x	$a_{x,B}=\gamma_{x,B}x_B$	1	8
ionic strength,				
molality basis	I_m, I	$I_m=\frac{1}{2}\Sigma m_B z_B{}^2$	mol kg^{-1}	
concentration basis	I_c, I	$I_c=\frac{1}{2}\Sigma c_B z_B{}^2$	mol m^{-3}	
osmotic coefficient,				
molality basis	ϕ_m	$\phi_m=(\mu_A{}^*-\mu_A)/(RTM_A\Sigma m_B)$	1	
mole fraction basis	ϕ_x	$\phi_x=(\mu_A-\mu_A{}^*)/(RT\ln x_A)$	1	
osmotic pressure	Π	$\Pi=c_B RT$ (ideal dilute solution)	Pa	

(14) An equivalent definition is $a_B=\lambda_B/\lambda_B{}^*$. The symbol* denotes pure substance. The symbol ° should not be used with this meaning, although it has been so used in the past.
(15) In the defining equations given here the pressure dependence of the activity has been neglected, as is often done for condensed phases at atmospheric pressure.
(16) An equivalent definition is $a_B=\lambda_B/\lambda_B{}^\circ$, where $\lambda_B{}^\circ=\exp(\mu_B{}^\circ/RT)$. The definition of μ° is different for a molality basis, a concentration basis, or a mole fraction basis; see p.47 below.

Other symbols and conventions in chemical thermodynamics

A more extensive description of this subject can be found in [1.j].

(i) *Symbols used as subscripts to denote a chemical process or reaction*
These symbols should be printed in roman (upright) type, without a full stop (period).

vaporization, evaporation (liquid→gas)	vap
sublimation (solid→gas)	sub
melting, fusion (solid→liquid)	fus
transition (between two phases)	trs
mixing of fluids	mix
solution (of solute in solvent)	sol
dilution (of a solution)	dil
adsorption	ads
displacement	dpl
immersion	imm
reaction in general	r
atomization	at
combustion reaction	c
formation reaction	f

(ii) *Recommended superscripts*

standard	$\ominus$, o
pure substance	*
infinite dilution	∞
ideal	id
activated complex, transition state	‡
excess quantity	E

(iii) *Examples of the use of these symbols*

The subscripts used to denote a chemical process, listed under (i) above, should be used as subscripts to the Δ symbol to denote the change in an extensive thermodynamic quantity associated with the process.

Example $\Delta_{\mathrm{vap}}H = H(\mathrm{g}) - H(\mathrm{l})$, for the enthalpy of vaporization, an extensive quantity proportional to the amount of substance vaporized.

The more useful quantity is usually the change divided by the amount of substance transferred, which should be denoted with an additional subscript m.

Example $\Delta_{\mathrm{vap}}H_{\mathrm{m}}$ for the molar enthalpy of vapourization.

However the subscript m is frequently omitted, particularly when the reader may tell from the units that a molar quantity is implied.

Example $\Delta_{\mathrm{vap}}H = 40.7\ \mathrm{kJ\ mol^{-1}}$ for H_2O at 373.15 K and 1 atm.

The subscript specifying the change is also sometimes attached to the symbol for the quantity rather than the Δ, so that the above quantity is denoted $\Delta H_{\mathrm{vap,m}}$ or simply ΔH_{vap}, but this is not recommended.

The subscript r is used to denote changes associated with a *chemical reaction*. Although symbols such as $\Delta_{\mathrm{r}}H$ should denote the integral enthalpy of reaction, $\Delta_{\mathrm{r}}H = H(\xi_2) - H(\xi_1)$, in practice this symbol is usually used to denote the change divided by the amount transferred, i.e. the change per extent of reaction, defined by the equation

$$\Delta_{\mathrm{r}}H = \sum_{\mathrm{B}} \nu_{\mathrm{B}} H_{\mathrm{B}} = (\partial H/\partial \xi)_{T,p}$$

It is thus essential to specify the stoichiometric reaction equation when giving numerical values for such quantities in order to define the extent of reaction ξ and the values of the stoichiometric numbers ν_{B}.

Example $N_2(g)+3H_2(g)=2NH_3(g)$, $\Delta_r H^{\circ}=-92.4\ \text{kJ mol}^{-1}$
$\Delta_r S^{\circ}=-199\ \text{J mol}^{-1}\ \text{K}^{-1}$

The mol^{-1} in the units identifies the quantities in this example as the change per extent of reaction. They may be called the molar enthalpy and entropy of reaction, and a subscript m may be added to the symbol, to emphasize the difference from the integral quantities if required.

The *standard reaction quantities* are particulary important. They are defined by the equations

$$\Delta_r H^{\circ} \quad (=\Delta_r H_m{}^{\circ}=\Delta H_m{}^{\circ})=\sum_B \nu_B H_B{}^{\circ}$$

$$\Delta_r S^{\circ} \quad (=\Delta_r S_m{}^{\circ}=\Delta S_m{}^{\circ})=\sum_B \nu_B S_B{}^{\circ}$$

$$\Delta_r G^{\circ} \quad (=\Delta_r G_m{}^{\circ}=\Delta G_m{}^{\circ})=\sum_B \nu_B \mu_B{}^{\circ}$$

The symbols in parentheses are alternatives. In view of the variety of styles in current use it is important to specify notation with care for these symbols. The relation to the affinity of the reaction is

$$-A=\Delta_r G=\Delta_r G^{\circ}+RT\ln\left(\prod_B a_B{}^{\nu_B}\right),$$

and the relation to the standard equilibrium constant is $\Delta_r G^{\circ}=-RT\ln K^{\circ}$.

The term *combustion* and symbol c denotes the complete oxidation of a substance. For the definition of complete oxidation of substances containing elements other than C, H and O see [45]. The corresponding reaction equation is written so that the stoichiometric number ν of the substance is -1.

Example The standard enthalpy of combustion of gaseous methane is $\Delta_c H^{\circ}(CH_4, \text{g}, 298.15\ \text{K})=-890.3\ \text{kJ mol}^{-1}$, implying the reaction $CH_4(g)+2O_2(g)\rightarrow CO_2(g)+2H_2O(l)$.

The term *formation* and symbol f denotes the formation of the substance from elements in their reference state (usually the most stable state of each element at the chosen temperature and pressure). The corresponding reaction equation is written so that the stoichiometric number ν of the substance is $+1$.

Example The standard entropy of formation of crystalline mercury II chloride is $\Delta_f S^{\circ}(HgCl_2, \text{cr}, 298.15\ \text{K})=-154.3\ \text{J mol}^{-1}\ \text{K}^{-1}$, implying the reaction $Hg(l)+Cl_2(g)\rightarrow HgCl_2(cr)$.

The term *atomization*, symbol at, denotes a process in which a substance is separated into its constituent atoms in the ground state in the gas phase. The corresponding reaction equation is written so that the stoichiometric number ν of the substance is -1.

Example The standard (internal) energy of atomization of liquid water is $\Delta_{at} U^{\circ}(H_2O, l)=625\ \text{kJ mol}^{-1}$ implying the reaction $H_2O(l)\rightarrow 2H(g)+O(g)$.

(iv) *Standard states* [1.j]

The standard chemical potential of substance B at temperature T, $\mu_B{}^{\circ}(T)$, is the value of the chemical potential under standard conditions, specified as follows. Three differently defined standard states are recognized.

For a gas phase. The standard state for a gaseous substance, whether pure or in a gaseous mixture, is the (hypothetical) state of the pure substance B in the gaseous phase at the standard pressure $p=p^{\circ}$ and exhibiting ideal gas behaviour. The standard chemical potential is defined as

$$\mu_B{}^{\circ}(T)=\lim_{p\to 0}\left[\mu_B(T, p, y_B, \ldots)-RT\ln(y_B p/p^{\circ})\right]$$

For a pure phase, or a mixture, or a solvent, in the liquid or solid state. The standard state for a liquid or solid substance, whether pure or in a mixture, or for a solvent, is the state of the pure substance B in the liquid or solid phase at the standard pressure $p=p^{\circ}$. The standard chemical potential is defined as

$$\mu_{\mathrm{B}}{}^{\circ}(T)=\mu_{\mathrm{B}}{}^{*}(T, p^{\circ})$$

For a solute in solution. For a solute in a liquid or solid solution the standard state is referenced to the ideal dilute behaviour of the solute. It is the (hypothetical) state of solute B at the standard molality m°, standard pressure p°, and exhibiting infinitely diluted solution behaviour. The standard chemical potential is defined as

$$\mu_{\mathrm{B}}{}^{\circ}(T)=[\mu_{\mathrm{B}}(T, p^{\circ}, m_{\mathrm{B}}, \ldots)-RT\ln(m_{\mathrm{B}}/m^{\circ})]^{\infty}.$$

The chemical potential of the solute B as a function of the molality m_{B} at constant pressure $p=p^{\circ}$ is then given by the expression

$$\mu_{\mathrm{B}}(m_{\mathrm{B}})=\mu_{\mathrm{B}}{}^{\circ}+RT\ln(m_{\mathrm{B}}\gamma_{m,\mathrm{B}}/m^{\circ}).$$

Sometimes (amount) concentration c is used as a variable in place of molality m; both of the above equations then have c in place of m throughout. Occasionally mole fraction x is used in place of m; both of the above equations then have x in place of m throughout, and $x^{\circ}=1$. Although the standard state of a solute is always referenced to ideal dilute behaviour, the definition of the standard state and the value of the standard chemical potential μ° are different depending on whether molality m, concentration c, or mole fraction x is used as a variable.

(v) *Standard pressures, molality, and concentration*

The standard pressure recommended by IUPAC since 1982 [1.j] is

$$p^{\circ}=10^{5}\ \mathrm{Pa}\quad(=1\ \mathrm{bar}),$$

and this is known as the standard state pressure.

Up to 1982 the standard pressure was usually taken to be

$$p^{\circ}=101\,325\ \mathrm{Pa}\ (=1\ \mathrm{atm},\ \text{called the 'standard atmosphere'}).$$

It may also sometimes be desirable to use a standard pressure substantially different from 10^{5} Pa, for example in tabulating data appropriate to high-pressure chemistry. It is therefore important always to specify the standard pressure adopted. The conversion of values corresponding to different p° is described in [46].

The standard molality is always taken as $m^{\circ}=1\ \mathrm{mol\,kg^{-1}}$.

The standard concentration is always taken as $c^{\circ}=1\ \mathrm{mol\,dm^{-3}}$.

In principle other m° and c° values could be used, but they would have to be specified.

(vi) *Values of thermodynamic quantities*

Values of many thermodynamic quantities represent basic chemical properties of substances and serve for further calculations. Extensive tabulations exist, e.g. [47, 48, 49]. Special care has to be taken in reporting the data and their uncertainties [23, 24].

2.12 CHEMICAL KINETICS

The recommendations given here are based on previous IUPAC recommendations [1.c,k and 25], which are not in complete agreement. For recommendations on reporting of chemical kinetic data see also [50].

Name	*Symbol*	*Definition*	*SI unit*	*Notes*
rate of change of quantity X	$\dot{X}$	$\dot{X}=\mathrm{d}X/\mathrm{d}t$	(varies)	1
rate of conversion	$\dot{\xi}$	$\dot{\xi}=\mathrm{d}\xi/\mathrm{d}t$	$\mathrm{mol\ s^{-1}}$	2
rate of concentration change (due to chemical reaction)	$r_\mathrm{B}, v_\mathrm{B}$	$r_\mathrm{B}=\mathrm{d}c_\mathrm{B}/\mathrm{d}t$	$\mathrm{mol\ m^{-3}\ s^{-1}}$	3, 4
rate of reaction (based on amount concentration)	v	$v=\dot{\xi}/V$ $=\nu_\mathrm{B}{}^{-1}\mathrm{d}c_\mathrm{B}/\mathrm{d}t$	$\mathrm{mol\ m^{-3}\ s^{-1}}$	2, 4
partial order of reaction	n_B	$v=k\Pi c_\mathrm{B}{}^{n_\mathrm{B}}$	1	5
overall order of reaction	n	$n=\Sigma n_\mathrm{B}$	1	
rate constant, rate coefficient	k	$v=k\Pi c_\mathrm{B}{}^{n_\mathrm{B}}$	$(\mathrm{mol^{-1}\,m^3})^{n-1}\mathrm{s^{-1}}$	
Boltzmann constant	k, k_B		$\mathrm{J\ K^{-1}}$	
half life	$t_{\frac{1}{2}}$	$c(t_{\frac{1}{2}})=c_0/2$	s	
relaxation time	τ	$\tau=1/(k_1+k_{-1})$	s	6
energy of activation, activation energy	E_a, E	$E_\mathrm{a}=RT^2\,\mathrm{d}\ln k/\mathrm{d}T$	$\mathrm{J\,mol^{-1}}$	
pre-exponential factor	A	$k=A\exp(-E_\mathrm{a}/RT)$	$(\mathrm{mol^{-1}\,m^3})^{n-1}\mathrm{s^{-1}}$	
volume of activation	$\Delta^{\ddagger}V$	$\Delta^{\ddagger}V=-RT\times$ $(\partial\ln k/\partial p)_T$	$\mathrm{m^3\,mol^{-1}}$	
collision diameter	d	$d_\mathrm{AB}=r_\mathrm{A}+r_\mathrm{B}$	m	
collision cross-section	σ	$\sigma_\mathrm{AB}=\pi d_\mathrm{AB}{}^2$	$\mathrm{m^2}$	
collision frequency	Z_A		$\mathrm{s^{-1}}$	7
collision number	$Z_\mathrm{AB}, Z_\mathrm{AA}$		$\mathrm{m^{-3}\,s^{-1}}$	7

(1) e.g. rate of change of pressure $\dot{p}=\mathrm{d}p/\mathrm{d}t$, for which the SI unit is $\mathrm{Pa\,s^{-1}}$.
(2) The reaction must be specified for which this quantity applies.
(3) The symbol and the definition apply to entities B.
(4) Note that r_B, v_B and v can also be defined on the basis of partial pressure, number concentration, surface concentration, etc., with analogous definitions. If necessary differently defined rates of reaction can be distinguished by a subscript, e.g. $v_p=\nu_\mathrm{B}{}^{-1}\,\mathrm{d}p_\mathrm{B}/\mathrm{d}t$, etc. Note that the rate of reaction can only be defined for a reaction of known and time-independent stoichiometry, in terms of a specified reaction equation; also the second equation for the rate of reaction follows from the first only if the volume V is constant. The derivatives must be those due to the chemical reaction considered; in open systems, such as flow systems, effects due to input and output processes must also be taken into account.
(5) The symbol applies to reactant B.
(6) The definition applies to reactions of first order in forward and backward directions with rate constants k_1 and k_{-1} respectively.
(7) Z_A is the number of collisions per time experienced by a single particle with particles of type A. Z_AA or Z_AB is the total number of AA or AB collisions per time and volume in a system containing only A molecules, or containing two types of molecules A and B. Three-body collisions can be treated in a similar way.

Name	*Symbol*	*Definition*	*SI unit*	*Notes*
collision frequency factor	z_{AB}, z_{AA}	$z_{AB}=Z_{AB}/Lc_A c_B$	$m^3\ mol^{-1}\ s^{-1}$	7
standard enthalpy of activation	$\Delta^{\ddagger}H^{\circ}, \Delta H^{\ddagger}$		$J\ mol^{-1}$	8
standard entropy of activation	$\Delta^{\ddagger}S^{\circ}, \Delta S^{\ddagger}$		$J\ mol^{-1}\ K^{-1}$	8
standard Gibbs energy of activation	$\Delta^{\ddagger}G^{\circ}, \Delta G^{\ddagger}$		$J\ mol^{-1}$	8
quantum yield, photochemical yield	ϕ		1	9

(8) The quantities $\Delta^{\ddagger}H^{\circ}$, $\Delta^{\ddagger}S^{\circ}$ and $\Delta^{\ddagger}G^{\circ}$ are used in the transition state theory of chemical reaction. They are normally used only in connection with elementary reactions. The relation between the rate constant k and these quantities is

$$k=\kappa(k_B T/h)\exp(-\Delta^{\ddagger}G^{\circ}/RT),$$

where k has the dimensions of a first-order rate constant and is obtained by multiplication of an nth-order rate constant by $(c^{\circ})^{n-1}$. κ is a transmission coefficient, and $\Delta^{\ddagger}G^{\circ}=\Delta^{\ddagger}H^{\circ}-T\Delta^{\ddagger}S^{\circ}$. Unfortunately the standard symbol $^{\circ}$ is usually omitted, and these quantities are usually written $\Delta H^{\ddagger}$, $\Delta S^{\ddagger}$ and $\Delta G^{\ddagger}$.

(9) The quantum yield ϕ is defined by the equation

$$\phi=\frac{\text{number of product molecules formed}}{\text{number of quanta absorbed}}.$$

2.13 ELECTROCHEMISTRY

Electrochemical concepts, terminology and symbols are more extensively described in [1.i].

Name	*Symbol*	*Definition*	*SI unit*	*Notes*
elementary charge (proton charge)	e		C	
Faraday constant	F	$F = eL$	$\mathrm{C\,mol^{-1}}$	
charge number of an ion	z	$z_B = Q_B/e$	1	1
ionic strength	I_c, I	$I_c = \frac{1}{2}\sum c_i z_i^2$	$\mathrm{mol\,m^{-3}}$	
mean ionic activity	$a_\pm$	$a_\pm = m_\pm \gamma_\pm / m^\ominus$	1	2, 3
mean ionic molality	$m_\pm$	$m_\pm^{(\nu_+ + \nu_-)} = m_+^{\nu_+} m_-^{\nu_-}$	$\mathrm{mol\,kg^{-1}}$	2
mean ionic activity coefficient	$\gamma_\pm$	$\gamma_\pm^{(\nu_+ + \nu_-)} = \gamma_+^{\nu_+} \gamma_-^{\nu_-}$	1	2
charge number of electrochemical cell reaction	n, (z)		1	4
electric potential difference (of a galvanic cell)	$\Delta V, E, U$	$\Delta V = V_R - V_L$	V	5
emf, electromotive force	E	$E = \lim_{I \to 0} \Delta V$	V	6
standard emf, standard potential of the electrochemical cell reaction	$E^\ominus$	$E^\ominus = -\Delta_r G^\ominus / nF = (RT/nF) \ln K^\ominus$	V	3, 7
standard electrode potential	$E^\ominus$		V	3, 8

(1) The definition applies to entities B.

(2) ν_+ and ν_- are the number of cations and anions per formula unit of an electrolyte $A_{\nu+}B_{\nu-}$.

Example for $Al_2(SO_4)_3$, $\nu_+ = 2$ and $\nu_- = 3$.

m_+ and m_-, and γ_+ and γ_-, are the separate cation and anion molalities and activity coefficients. If the molality of $A_{\nu+}B_{\nu-}$ is m, then $m_+ = \nu_+ m$ and $m_- = \nu_- m$. A similar definition is used on a concentration scale for the mean ionic concentration $c_\pm$.

(3) The symbol $^\ominus$ or $^\circ$ is used to indicate standard. They are equally acceptable.

(4) n is the number of electrons transferred according to the cell reaction (or half cell reactions) as written; n is a positive integer.

(5) V_R and V_L are the potentials of the electrodes shown on the right- and left-hand sides, respectively, in the diagram representing the cell. When ΔV is positive, positive charge flows from left to right through the cell, and from right to left in the external circuit, if the cell is short-circuited.

(6) The definition of emf is discussed on p.53. The symbol E_{MF} is no longer recommended for this quantity.

(7) $\Delta_r G^\ominus$ and $K^\ominus$ apply to the cell reaction in the direction in which reduction occurs at the right-hand electrode and oxidation at the left-hand electrode, in the diagram representing the cell (see p.53). (Note the mnemonic 'reduction at the right'.)

(8) Standard potential of an electrode reaction, abbreviated as standard electrode potential, is the value of the standard emf of a cell in which molecular hydrogen is oxidized to solvated protons at the left-hand electrode. For example, the standard potential of the Zn^{2+}/Zn electrode, denoted $E^\ominus(Zn^{2+}/Zn)$, is the emf of the cell in which the reaction $Zn^{2+}(aq) + H_2 \rightarrow 2H^+(aq) + Zn$ takes place under standard conditions (see p.54). The concept of an *absolute* electrode potential is discussed in reference [26].

Name	*Symbol*	*Definition*	*SI unit*	*Notes*
emf of the cell, potential of the electrochemical cell reaction	E	$E = E^{\ominus} - (RT/nF) \times \sum \nu_i \ln a_i$	V	9
pH	pH	$\mathrm{pH} \approx -\lg\left[\frac{c(\mathrm{H}^+)}{\mathrm{mol\,dm^{-3}}}\right]$	1	10
inner electric potential	ϕ	$\nabla\phi = -\boldsymbol{E}$	V	11
outer electric potential	ψ	$\psi = Q/4\pi\varepsilon_0 r$	V	12
surface electric potential	χ	$\chi = \phi - \psi$	V	
Galvani potential difference	$\Delta\phi$	$\Delta_\alpha^\beta\phi = \phi^\beta - \phi^\alpha$	V	13
volta potential difference	$\Delta\psi$	$\Delta_\alpha^\beta\psi = \psi^\beta - \psi^\alpha$	V	14
electrochemical potential	$\tilde{\mu}$	$\tilde{\mu}_\mathrm{B}{}^\alpha = (\partial G/\partial n_\mathrm{B}{}^\alpha)$	$\mathrm{J\,mol^{-1}}$	1, 15
electric current	I	$I = \mathrm{d}Q/\mathrm{d}t$	A	16
(electric) current density	j	$j = I/A$	$\mathrm{A\,m^{-2}}$	16
(surface) charge density	σ	$\sigma = Q/A$	$\mathrm{C\,m^{-2}}$	
electrode reaction rate constant	k	$k_\mathrm{ox} = I_\mathrm{a}/(nFA\prod_i c_i{}^{n_i})$	(varies)	17, 18
mass transfer coefficient, diffusion rate constant	k_d	$k_\mathrm{d,B} = \lvert\nu_\mathrm{B}\rvert I_\mathrm{l,B}/nFcA$	$\mathrm{m\,s^{-1}}$	1,18
thickness of diffusion layer	δ	$\delta_\mathrm{B} = D_\mathrm{B}/k_\mathrm{d,B}$	m	1
transfer coefficient (electrochemical)	α	$\alpha_\mathrm{c} = \frac{-\lvert\nu\rvert RT}{nF}\frac{\partial \ln \lvert I_\mathrm{c}\rvert}{\partial E}$	1	16, 18
overpotential	η	$\eta = E_I - E_{I=0} - IR_u$	V	18
electrokinetic potential (zeta potential)	ζ		V	

(9) $\sum \nu_i \ln a_i$ refers to the cell reaction, with ν_i positive for products and negative for reactants; for the complete cell reaction only mean ionic activities $a_\pm$ are involved.
(10) The precise definition of pH is discussed on p.54. The symbol pH is an exception to the general rules for the symbols of physical quantities (p.5) in that it is a two-letter symbol and it is always printed in roman (upright) type.
(11) $\boldsymbol{E}$ is the electric field strength within the phase concerned.
(12) The definition is an example specific to a conducting sphere of excess charge Q and radius r.
(13) $\Delta\phi$ is the electric potential difference between points within the bulk phases α and β; it is measurable only if the phases are of identical composition.
(14) $\Delta\psi$ is the electric potential difference due to the charge on phases α and β. It is measurable or calculable by classical electrostatics from the charge distribution.
(15) The chemical potential is related to the electrochemical potential by the equation $\mu_\mathrm{B}{}^\alpha = \tilde{\mu}_\mathrm{B}{}^\alpha - z_\mathrm{B}F\phi^\alpha$. For an uncharged species, $z_\mathrm{B} = 0$, the electrochemical potential is equal to the chemical potential.
(16) I, j and α may carry one of the subscripts: a for anodic, c for cathodic, e or o for exchange, or l for limiting. I_a and I_c are the anodic and cathodic partial currents. The cathode is the electrode where reduction takes place, and the anode is the electrode where oxidation takes place.
(17) For reduction the rate constant k_red can be defined analogously in terms of the cathodic current I_c. For first-order reaction the SI unit is $\mathrm{m\,s^{-1}}$. n_i is the order of reaction with respect to component i.
(18) For more information on kinetics of electrode reactions and on transport phenomena in electrolyte systems see [27] and [28].

Name	*Symbol*	*Definition*	*SI unit*	*Notes*
conductivity	$\kappa, (\sigma)$	$\kappa = j/E$	$S\,m^{-1}$	11, 19
conductivity cell constant	K_{cell}	$K_{cell} = \kappa R$	m^{-1}	
molar conductivity (of an electrolyte)	Λ	$\Lambda_B = \kappa/c_B$	$S\,m^2\,mol^{-1}$	1
ionic conductivity, molar conductivity of an ion	λ	$\lambda_B = \lvert z_B \rvert F u_B$	$S\,m^2\,mol^{-1}$	1, 20
electric mobility	$u, (\mu)$	$u_B = v_B/E$	$m^2\,V^{-1}\,s^{-1}$	1, 11
transport number	t	$t_B = j_B/\Sigma j_i$	1	1
reciprocal radius of ionic atmosphere	κ	$\kappa = (2F^2 I/\varepsilon RT)^{\frac{1}{2}}$	m^{-1}	21

(19) Conductivity was formerly called specific conductance.
(20) It is important to specify the entity to which molar conductivity refers; thus for example $\lambda(Mg^{2+}) = 2\lambda(\frac{1}{2}Mg^{2+})$. It is standard practice to choose the entity to be $1/z_B$ of an ion of charge number z_B, so that for example molar conductivities for potassium, barium and lanthanum ions would be quoted as $\lambda(K^+)$, $\lambda(\frac{1}{2}Ba^{2+})$, or $\lambda(1/3\,La^{3+})$.
(21) κ appears in Debye–Hückel theory. $\kappa^{-1} = L_D$, the Debye length, which appears in Debye–Hückel theory, Gouy–Chapman theory, and the theory of semiconductor space charge.

Conventions concerning the signs of electric potential differences, electromotive forces, and electrode potentials[1]

(i) *The electric potential difference for a galvanic cell*
The cell should be represented by a diagram, for example:

$$Zn|Zn^{2+} \,\vdots\, Cu^{2+}|Cu.$$

A single vertical bar (|) should be used to represent a phase boundary, a dashed vertical bar (¦) to represent a junction between miscible liquids, and double dashed vertical bars (¦¦) to represent a liquid junction in which the liquid junction potential is assumed to be eliminated. The electric potential difference, denoted ΔV or E, is equal in sign and magnitude to the electric potential of a metallic conducting lead on the right minus that of a similar lead on the left. The emf (electromotive force), also usually denoted E, is the limiting value of the electric potential difference for zero current through the cell, all local charge transfer equilibria and chemical equilibria being established. Note that the symbol E is often used for both the potential difference and the emf, and this can sometimes lead to confusion.

When the reaction of the cell is written as

$$\tfrac{1}{2}Zn + \tfrac{1}{2}Cu^{2+} \rightarrow \tfrac{1}{2}Zn^{2+} + \tfrac{1}{2}Cu, \qquad n = 1$$

or

$$Zn + Cu^{2+} \rightarrow Zn^{2+} + Cu, \qquad n = 2,$$

this implies a cell diagram drawn, as above, so that this reaction takes place when positive electricity flows through the cell from left to right (and therefore through the outer part of the circuit from right to left). In the above example the right-hand electrode is positive (unless the ratio $[Cu^{2+}]/[Zn^{2+}]$ is

(1) These are in accordance with the 'Stockholm Convention' of 1953.

extremely small), so that this is the direction of spontaneous flow if a wire is connected across the two electrodes. If however the reaction is written as

$$\tfrac{1}{2}\mathrm{Cu}+\tfrac{1}{2}\mathrm{Zn}^{2+}\rightarrow\tfrac{1}{2}\mathrm{Cu}^{2+}+\tfrac{1}{2}\mathrm{Zn}, \qquad n=1$$

or

$$\mathrm{Cu}+\mathrm{Zn}^{2+}\rightarrow\mathrm{Cu}^{2+}+\mathrm{Zn}, \qquad n=2,$$

this implies the cell diagram

$$\mathrm{Cu}\,|\,\mathrm{Cu}^{2+} \;\vdots\; \mathrm{Zn}^{2+}\,|\,\mathrm{Zn}$$

and the electric potential difference of the cell so specified will be negative. Thus a cell diagram may be drawn either way round, and correspondingly the electric potential difference appropriate to the diagram may be either positive or negative.

(ii) *Electrode potential* (*potential of an electrode reaction*)

The so-called electrode potential of an electrode is defined as the emf of a cell in which the electrode on the left is a standard hydrogen electrode and the electrode on the right is the electrode in question. For example, for the silver/silver chloride electrode (written $\mathrm{Cl^-(aq)|AgCl|Ag}$) the cell in question is

$$\mathrm{Pt}\,|\,\mathrm{H_2}(\mathrm{g}, p=p^\circ)\,|\,\mathrm{HCl}\ (\mathrm{aq}, a_\pm=1) \;\vdots\; \mathrm{HCl}\ (\mathrm{aq}, a_\pm')\,|\,\mathrm{AgCl}\,|\,\mathrm{Ag}$$

A liquid junction will be necessary in this cell whenever $a_\pm'$(HCl) on the right differs from $a_\pm$(HCl) on the left. The reaction taking place at the silver/silver chloride electrode is

$$\mathrm{AgCl(s)}+\mathrm{e^-}\rightarrow\mathrm{Ag(s)}+\mathrm{Cl^-(aq)}$$

The complete cell reaction is

$$\mathrm{AgCl(s)}+\tfrac{1}{2}\mathrm{H_2(g)}\rightarrow\mathrm{H^+(aq)}+\mathrm{Cl^-(aq)}+\mathrm{Ag(s)}$$

In the standard state of the hydrogen electrode, $p(\mathrm{H_2})=p^\circ=10^5$ Pa and $a_\pm(\mathrm{HCl})=1$, the emf of this cell is the electrode potential of the silver/silver chloride electrode. If, in addition, the mean activity of the HCl in the silver/silver chloride electrode $a_\pm(\mathrm{HCl})=1$, then the emf is equal to E° for this electrode. The standard electrode potential for $\mathrm{HCl(aq)|AgCl|Ag}$ has the value $E^\circ=+0.22217$ V at 298.15 K. For $p^\circ=101325$ Pa the standard potential of this electrode (and of any electrode involving only condensed phases) is higher by 0.17 mV; i.e.

$$E^\circ(101\,325\ \mathrm{Pa})=E^\circ(10^5\ \mathrm{Pa})+0.17\ \mathrm{mV}$$

A compilation of standard electrode potentials, and their conversion between different standard pressures, can be found in [29]. Notice that in writing the cell whose emf represents an electrode potential, it is important that the hydrogen electrode should always be on the left.

(iii) *Operational definition of* pH [30]

The notional definition of pH given in the table above is in practice replaced by the following operational definition. For a solution X the emf E(X) of the galvanic cell

reference electrode	KCl (aq, $m>3.5\ \mathrm{mol\ kg^{-1}}$)	⋮	solution X	$\mathrm{H_2(g)}$	Pt

is measured, and likewise the emf E(S) of the cell that differs only by the replacement of the solution X of unknown pH(X) by the solution S of standard pH(S). The unknown pH is then given by

$$\mathrm{pH(X)}=\mathrm{pH(S)}+(E_\mathrm{S}-E_\mathrm{X})F/(RT\ln 10)$$

Thus defined, pH is dimensionless. Values of pH(S) for several standard solutions and temperatures are listed in [30]. The reference pH standard is an aqueous solution of potassium hydrogen phthalate at a molality of exactly 0.05 mol kg^{-1}: at 25 °C (298.15 K) this has a pH of 4.005.

In practice a glass electrode is almost always used in place of the Pt|H_2 electrode. The cell might then take the form

$$\begin{array}{l|l} \text{reference} & \text{KCl(aq,} \\ \text{electrode} & m > 3.5\ \text{mol kg}^{-1}) \end{array} \;\vdots\vdots\; \text{solution X} \mid \text{glass} \mid \text{H}^+, \text{Cl}^- \mid \text{AgCl} \mid \text{Ag}$$

The solution to the right of the glass electrode is usually a buffer solution of KH_2PO_4 and Na_2HPO_4, with 0.1 mol dm^{-3} of NaCl. The reference electrode is usually a calomel electrode, silver/silver chloride electrode, or a thallium amalgam/thallous chloride electrode. The emf of this cell depends on $a(H^+)$ in the solution X in the same way as that of the cell with the Pt|H_2 electrode, and thus the same procedure is followed.

In the restricted range of dilute aqueous solutions having amount concentrations less than 0.1 mol dm^{-3} and being neither strongly acidic nor strongly alkaline ($2 < \text{pH} < 12$) the above definition is such that

$$\begin{aligned} \text{pH} &= -\lg\,[\gamma_{\pm}\, c(\text{H}^+)/(\text{mol dm}^{-3})] \pm 0.02, \\ &= -\lg\,[\gamma_{\pm}\, m(\text{H}^+)/(\text{mol kg}^{-1})] \pm 0.02, \end{aligned}$$

where $c(H^+)$ denotes the amount concentration of hydrogen ion H^+ and $m(H^+)$ the corresponding molality, and $\gamma_{\pm}$ denotes the mean ionic activity coefficient of a typical uni-univalent electrolyte in the solution on a concentration basis or a molality basis as appropriate. For further information on the definition of pH see [30].

2.14 COLLOID AND SURFACE CHEMISTRY

The recommendations given here are based on more extensive IUPAC recommendations [1.e–h] and [30a].

Name	*Symbol*	*Definition*	*SI unit*	*Notes*
specific surface area	a, a_s, s	$a = A/m$	$m^2\ kg^{-1}$	
surface amount of B, adsorbed amount of B	n_B^s, n_B^a		mol	1
surface excess of B	n_B^σ		mol	2
surface excess concentration of B	Γ_B, (Γ_B^σ)	$\Gamma_B = n_B^\sigma/A$	$mol\ m^{-2}$	2
total surface excess concentration	Γ, (Γ^σ)	$\Gamma = \sum_i \Gamma_i$	$mol\ m^{-2}$	
area per molecule	a, σ	$a_B = A/N_B^\sigma$	m^2	3
area per molecule in a filled monolayer	a_m, σ_m	$a_{m,B} = A/N_{m,B}$	m^2	3
surface coverage	θ	$\theta = N_B^\sigma/N_{m,B}$	1	3
contact angle	θ		1, rad	
film thickness	t, h, δ		m	
thickness of (surface or interfacial) layer	τ, δ, t		m	
surface tension, interfacial tension	γ, σ	$\gamma = (\partial G/\partial A_s)_{T,p}$	$N\ m^{-1}$, $J\ m^{-2}$	
film tension	Σ_f	$\Sigma_f = 2\gamma_f$	$N\ m^{-1}$	4
reciprocal thickness of the double layer	κ	$\kappa = [2F^2 I_c/\varepsilon RT]^{\frac{1}{2}}$	m^{-1}	
average molar masses				
number-average	M_n	$M_n = \Sigma n_i M_i/\Sigma n_i$	$kg\ mol^{-1}$	
mass-average	M_m	$M_m = \Sigma n_i M_i^2/\Sigma n_i M_i$	$kg\ mol^{-1}$	
Z-average	M_Z	$M_Z = \Sigma n_i M_i^3/\Sigma n_i M_i^2$	$kg\ mol^{-1}$	
sedimentation coefficient	s	$s = v/a$	s	5
van der Waals constant	λ		J	
retarded van der Waals constant	β, B		J	
van der Waals–Hamaker constant	A_H		J	
surface pressure	π^s, π	$\pi^s = \gamma^0 - \gamma$	$N\ m^{-1}$	6

(1) The value of n_B^s depends on the thickness assigned to the surface layer.

(2) The values of n_B^σ and Γ_B depend on the convention used to define the position of the Gibbs surface. They are given by the excess amount of B or surface concentration of B over values that would apply if each of the two bulk phases were homogeneous right up to the Gibbs surface. See [1.e], and also additional recommendations on p.57.

(3) N_B^σ is the number of adsorbed molecules ($N_B^\sigma = L n_B^\sigma$), and N_B^m is the number of adsorbed molecules in a filled monolayer. The definition to entities B.

(4) The definition applies only to a symmetrical film, for which the two bulk phases on either side of the film are the same, and γ_f is the surface tension of a film/bulk interface.

(5) In the definition, v is the velocity of sedimentation and a is the acceleration of free fall or centrifugation. The symbol for a limiting sedimentation coefficient is $[s]$, for a reduced sedimentation coefficient s°, and for a reduced limiting sedimentation coefficient $[s^\circ]$; see [1.e] for further details.

(6) In the definition, γ^0 is the surface tension of the clean surface and γ that of the covered surface.

Additional recommendations

The superscript s denotes the properties of a surface or interfacial layer. In the presence of adsorption it may be replaced by the superscript a.

Examples	Helmholtz energy of interfacial layer	A^s
	amount of adsorbed substance	n^a, n^s
	amount of adsorbed O_2	$n^a(O_2)$, $n^s(O_2)$, or $n(O_2, a)$

The subscript m denotes the properties of a monolayer.

Example	area per molecule B in a monolayer	$a_m(B)$

The superscript σ is used to denote a surface excess property relative to the Gibbs surface.

Example	surface excess amount (or Gibbs surface excess of B)	n_B^σ

In general the values of Γ_A and Γ_B depend on the position chosen for the Gibbs dividing surface. However, two quantities, $\Gamma_B^{(A)}$ and $\Gamma_B^{(n)}$ (and correspondingly $n_B^{\sigma(A)}$ and $n_B^{\sigma(n)}$), may be defined in a way that is invariant to this choice (see [1.e]). $\Gamma_B^{(A)}$ is called the *relative* surface excess concentration of B with respect to A, or more simply the relative adsorption of B; it is the value of Γ_B when the surface is chosen to make $\Gamma_A = 0$. $\Gamma_B^{(n)}$ is called the *reduced* surface excess concentration of B, or more simply the reduced adsorption of B; it is the value of Γ_B when the surface is chosen to make the total excess $\Gamma = \sum_i \Gamma_i = 0$.

Properties of phases (α, β, γ) may be denoted by corresponding superscript indices.

Examples	surface tension of phase α	γ^α
	interfacial tension between phases α and β	$\gamma^{\alpha\beta}$

Symbols of thermodynamic quantities divided by surface area are usually the corresponding lower case letters; an alternative is to use a circumflex.

Example	interfacial entropy per area	$s^s(=\hat{s}^s) = S^s/A$

The following abbreviations are used in colloid chemistry:

c.c.c.	critical coagulation concentration
c.m.c.	critical micellization concentration
i.e.p.	isoelectric point
p.z.c.	point of zero charge

2.15 TRANSPORT PROPERTIES

The names and symbols recommended here are in agreement with those recommended by IUPAP [3] and ISO [4.m]. Further information on transport phenomena in electrochemical systems can also be found in [27].

Name	*Symbol*	*Definition*	*SI units*	*Notes*
flux (of a quantity X)	J_X, J	$J_X = A^{-1}\, \mathrm{d}X/\mathrm{d}t$	(varies)	
volume flow rate	q_V, $\dot{V}$	$q_v = \mathrm{d}V/\mathrm{d}t$	$\mathrm{m^3\ s^{-1}}$	
mass flow rate	q_m, $\dot{m}$	$q_m = \mathrm{d}m/\mathrm{d}t$	$\mathrm{kg\ s^{-1}}$	
mass transfer coefficient	k_d		$\mathrm{m\ s^{-1}}$	
heat flow rate	ϕ	$\phi = \mathrm{d}q/\mathrm{d}t$	W	
heat flux	J_q	$J_q = \phi/A$	$\mathrm{W\ m^{-2}}$	
thermal conductance	G	$G = \phi/\Delta T$	$\mathrm{W\ K^{-1}}$	
thermal resistance	R	$R = 1/G$	$\mathrm{K\ W^{-1}}$	
thermal conductivity	λ, k	$\lambda = J_q/(\mathrm{d}T/\mathrm{d}l)$	$\mathrm{W\ m^{-1}\ K^{-1}}$	
coefficient of heat transfer	h, (k, K, α)	$h = J_q/\Delta T$	$\mathrm{W\ m^{-2}\ K^{-1}}$	
thermal diffusivity	a	$a = \lambda/\rho c_p$	$\mathrm{m^2\ s^{-1}}$	
diffusion coefficient	D	$D = J_n/(\mathrm{d}c/\mathrm{d}l)$	$\mathrm{m^2\ s^{-1}}$	

The following symbols are used in the definitions of the dimensionless quantities: mass (m), time (t), volume (V), area (A), density (ρ), speed (v), length (l), viscosity (η), pressure (p), acceleration of free fall (g), cubic expansion coefficient (α), temperature (T), surface tension (γ), speed of sound (c), mean free path (λ), frequency (f), thermal diffusivity (a), coefficient of heat transfer (h), thermal conductivity (k), specific heat capacity at constant pressure (c_p), diffusion coefficient (D), mole fraction (x), mass transfer coefficient (k_d), permeability (μ), electric conductivity (κ), and magnetic flux density (B).

Name	*Symbol*	*Definition*	*SI unit*	*Notes*
Reynolds number	Re	$Re = \rho v l/\eta$	1	
Euler number	Eu	$Eu = \Delta p/\rho v^2$	1	
Froude number	Fr	$Fr = v/(lg)^{\frac{1}{2}}$	1	
Grashof number	Gr	$Gr = l^3 g\alpha\Delta T\rho^2/\eta^2$	1	
Weber number	We	$We = \rho v^2 l/\gamma$	1	
Mach number	Ma	$Ma = v/c$	1	
Knudsen number	Kn	$Kn = \lambda/l$	1	
Strouhal number	Sr	$Sr = lf/v$	1	
Fourier number	Fo	$Fo = at/l^2$	1	
Péclet number	Pe	$Pe = vl/a$	1	
Rayleigh number	Ra	$Ra = l^3 g\alpha\Delta T\rho/\eta a$	1	
Nusselt number	Nu	$Nu = hl/k$	1	
Stanton number	St	$St = h/\rho v c_p$	1	
Fourier number for mass transfer	Fo^*	$Fo^* = Dt/l^2$	1	1
Péclet number for mass transfer	Pe^*	$Pe^* = vl/D$	1	1
Grashof number for mass transfer	Gr^*	$Gr^* = l^3 g\left(\frac{\partial\rho}{\partial x}\right)_{T,p}\left(\frac{\Delta x\rho}{\eta}\right)$	1	1

(1) This quantity applies to the transport of matter in binary mixtures.

Name	Symbol	Definition	SI unit	Notes
Nusselt number for mass transfer	Nu^*	$Nu^* = k_d l/D$	1	1, 2
Stanton number for mass transfer	St^*	$St^* = k_d/v$	1	1
Prandtl number	Pr	$Pr = \eta/\rho a$	1	
Schmidt number	Sc	$Sc = \eta/\rho D$	1	
Lewis number	Le	$Le = a/D$	1	
magnetic Reynolds number	Rm, Re_m	$Rm = v\mu\kappa l$	1	
Alfvén number	Al	$Al = v(\rho\mu)^{\frac{1}{2}}/B$	1	
Hartmann number	Ha	$Ha = Bl(\kappa/\eta)^{\frac{1}{2}}$	1	
Cowling number	Co	$Co = B^2/\mu\rho v^2$	1	

(2) The name Sherwood number and symbol *Sh* has been widely used for this quantity.

3
Definitions and symbols for units

3.1 THE INTERNATIONAL SYSTEM OF UNITS (SI)

The International System of units (SI) was adopted by the 11th General Conference on Weights and Measures (CGPM) in 1960 [2]. It is a coherent system of units built from seven *SI base units*, one for each of the seven dimensionally independent base quantities (see section 1.2): they are the metre, kilogram, second, ampere, kelvin, mole, and candela, for the dimensions length, mass, time, electric current, thermodynamic temperature, amount of substance, and luminous intensity, respectively. The definitions of the SI base units are given in section 3.2. The *SI derived units* are expressed as products of powers of the base units, analogous to the corresponding relations between physical quantities but with numerical factors equal to unity [2].

In the International System there is only one SI unit for each physical quantity. This is either the appropriate SI base unit itself (see table 3.3) or the appropriate SI derived unit (see tables 3.4 and 3.5). However, any of the approved decimal prefixes, called *SI prefixes*, may be used to construct decimal multiples or submultiples of SI units (see table 3.6).

It is recommended that only SI units be used in science and technology (with SI prefixes where appropriate). Where there are special reasons for making an exception to this rule, it is recommended always to define the units used in terms of SI units.

3.2 DEFINITIONS OF THE SI BASE UNITS [2]

metre: The metre is the length of path travelled by light in vacuum during a time interval of 1/299 792 458 of a second (17th CGPM, 1983).

kilogram: The kilogram is the unit of mass; it is equal to the mass of the international prototype of the kilogram (3rd CGPM, 1901).

second: The second is the duration of 9 192 631 770 periods of the radiation corresponding to the transition between the two hyperfine levels of the ground state of the caesium-133 atom (13th CGPM, 1967).

ampere: The ampere is that constant current which, if maintained in two straight parallel conductors of infinite length, of negligible circular cross-section, and placed 1 metre apart in vacuum, would produce between these conductors a force equal to 2×10^{-7} newton per metre of length (9th CGPM, 1948).

kelvin: The kelvin, unit of thermodynamic temperature, is the fraction 1/273.16 of the thermodynamic temperature of the triple point of water (13th CGPM, 1967).

mole: The mole is the amount of substance of a system which contains as many elementary entities as there are atoms in 0.012 kilogram of carbon-12. When the mole is used, the elementary entities must be specified and may be atoms, molecules, ions, electrons, other particles, or specified groups of such particles (14th CGPM, 1971).

Examples of the use of the mole

1 mol of H_2 contains about 6.022×10^{23} H_2 molecules, or 12.044×10^{23} H atoms
1 mol of HgCl has a mass of 236.04 g
1 mol of Hg_2Cl_2 has a mass of 472.08 g
1 mol of $Hg_2{}^{2+}$ has a mass of 401.18 g and a charge of 192.97 kC
1 mol of $Fe_{0.91}S$ has a mass of 82.88 g
1 mol of e^- has a mass of 548.60 μg and a charge of −96.49 kC
1 mol of photons whose frequency is 10^{14} Hz has energy of about 39.90 kJ

See also section 2.10, p.41.

candela: The candela is the luminous intensity, in a given direction, of a source that emits monochromatic radiation of frequency 540×10^{12} hertz and that has a radiant intensity in that direction of (1/683) watt per steradian (16th CGPM, 1979).

3.3 NAMES AND SYMBOLS FOR THE SI BASE UNITS

The symbols listed here are internationally agreed and should not be changed in other languages or scripts. See sections 1.3 and 1.4 on the printing of symbols for units. Recommended representations for these symbols for use in systems with limited character sets can be found in [6].

Physical quantity	*Name of SI unit*	*Symbol for SI unit*
length	metre	m
mass	kilogram	kg
time	second	s
electric current	ampere	A
thermodynamic temperature	kelvin	K
amount of substance	mole	mol
luminous intensity	candela	cd

3.4 SI DERIVED UNITS WITH SPECIAL NAMES AND SYMBOLS

Physical quantity	*Name of SI unit*	*Symbol for SI unit*	*Expression in terms of SI base units*	
frequency[1]	hertz	Hz	s^{-1}	
force	newton	N	$m\ kg\ s^{-2}$	
pressure, stress	pascal	Pa	$N\ m^{-2}$	$=m^{-1}\ kg\ s^{-2}$
energy, work, heat	joule	J	$N\ m$	$=m^2\ kg\ s^{-2}$
power, radiant flux	watt	W	$J\ s^{-1}$	$=m^2\ kg\ s^{-3}$
electric charge	coulomb	C	$A\ s$	
electric potential, electromotive force	volt	V	$J\ C^{-1}$	$=m^2\ kg\ s^{-3}\ A^{-1}$
electric resistance	ohm	Ω	$V\ A^{-1}$	$=m^2\ kg\ s^{-3}\ A^{-2}$
electric conductance	siemens	S	Ω^{-1}	$=m^{-2}\ kg^{-1}\ s^3\ A^2$
electric capacitance	farad	F	$C\ V^{-1}$	$=m^{-2}\ kg^{-1}\ s^4\ A^2$
magnetic flux density	tesla	T	$V\ s\ m^{-2}$	$=kg\ s^{-2}\ A^{-1}$
magnetic flux	weber	Wb	$V\ s$	$=m^2\ kg\ s^{-2}\ A^{-1}$
inductance	henry	H	$V\ A^{-1}\ s$	$=m^2\ kg\ s^{-2}\ A^{-2}$
Celsius temperature[2]	degree Celsius	°C	K	
luminous flux	lumen	lm	cd sr	
illuminance	lux	lx	$cd\ sr\ m^{-2}$	
activity[3] (radioactive)	becquerel	Bq	s^{-1}	
absorbed dose[3] (of radiation)	gray	Gy	$J\ kg^{-1}$	$=m^2\ s^{-2}$
dose equivalent[3] (dose equivalent index)	sievert	Sv	$J\ kg^{-1}$	$=m^2\ s^{-2}$
plane angle[4]	radian	rad	1	$=m\ m^{-1}$
solid angle[4]	steradian	sr	1	$=m^2\ m^{-2}$

(1) For radial (circular) frequency and for angular velocity the unit rad s^{-1}, or simply s^{-1}, should be used, and this may *not* be simplified to Hz. The unit Hz should be used *only* for frequency in the sense of cycles per second.
(2) The Celsius temperature θ is defined by the equation:

$$\theta/°C = T/K - 273.15.$$

The SI unit of Celsius temperature interval is the degree Celsius, °C, which is equal to the kelvin, K. °C should be treated as a single symbol, with no space between the ° sign and the letter C. (The symbol °K, and the symbol °, should no longer be used.)
(3) The units gray and sievert are admitted for reasons of safeguarding human health [2].
(4) The units radian and steradian are described as 'SI supplementary units' [2]. However, in chemistry, as well as in physics [3], they are usually treated as dimensionless derived units, and this was recognized by CIPM in 1980. Since they are then of dimension 1, this leaves open the possibility of including them or omitting them in expressions of SI derived units. In practice this means that rad and sr may be used when appropriate and may be omitted if clarity is not lost thereby.

3.5 SI DERIVED UNITS FOR OTHER QUANTITIES

This table gives examples of other SI derived units; the list is merely illustrative.

Physical quantity	*Expression in terms of SI base units*	
area	m^2	
volume	m^3	
speed, velocity	$m\ s^{-1}$	
angular velocity	s^{-1}, $rad\ s^{-1}$	
acceleration	$m\ s^{-2}$	
moment of force	$N\ m$	$= m^2\ kg\ s^{-2}$
wavenumber	m^{-1}	
density, mass density	$kg\ m^{-3}$	
specific volume	$m^3\ kg^{-1}$	
amount concentration[1]	$mol\ m^{-3}$	
molar volume	$m^3\ mol^{-1}$	
heat capacity, entropy	$J\ K^{-1}$	$= m^2\ kg\ s^{-2}\ K^{-1}$
molar heat capacity, molar entropy	$J\ K^{-1}\ mol^{-1}$	$= m^2\ kg\ s^{-2}\ K^{-1}\ mol^{-1}$
specific heat capacity, specific entropy	$J\ K^{-1}\ kg^{-1}$	$= m^2\ s^{-2}\ K^{-1}$
molar energy	$J\ mol^{-1}$	$= m^2\ kg\ s^{-2}\ mol^{-1}$
specific energy	$J\ kg^{-1}$	$= m^2\ s^{-2}$
energy density	$J\ m^{-3}$	$= m^{-1}\ kg\ s^{-2}$
surface tension	$N\ m^{-1}$, $J\ m^{-2}$	$= kg\ s^{-2}$
heat flux density, irradiance	$W\ m^{-2}$	$= kg\ s^{-3}$
thermal conductivity	$W\ m^{-1}\ K^{-1}$	$= m\ kg\ s^{-3}\ K^{-1}$
kinematic viscosity, diffusion coefficient	$m^2\ s^{-1}$	
dynamic viscosity	$N\ s\ m^{-2}$, $Pa\ s$	$= m^{-1}\ kg\ s^{-1}$
electric charge density	$C\ m^{-3}$	$= m^{-3}\ s\ A$
electric current density	$A\ m^{-2}$	
conductivity	$S\ m^{-1}$	$= m^{-3}\ kg^{-1}\ s^3\ A^2$
molar conductivity	$S\ m^2\ mol^{-1}$	$= kg^{-1}\ mol^{-1}\ s^3\ A^2$
permittivity	$F\ m^{-1}$	$= m^{-3}\ kg^{-1}\ s^4\ A^2$
permeability	$H\ m^{-1}$	$= m\ kg\ s^{-2}\ A^{-2}$
electric field strength	$V\ m^{-1}$	$= m\ kg\ s^{-3}\ A^{-1}$
magnetic field strength	$A\ m^{-1}$	
luminance	$cd\ m^{-2}$	
exposure (X and γ rays)	$C\ kg^{-1}$	$= kg^{-1}\ s\ A$
absorbed dose rate	$Gy\ s^{-1}$	$= m^2\ s^{-3}$

(1) The words 'amount concentration' are an abbreviation for 'amount-of-substance concentration'. When there is not likely to be any ambiguity this quantity may be called simply 'concentration'.

3.6 SI PREFIXES

To signify decimal multiples and submultiples of SI units the following prefixes may be used [2].

Submultiple	*Prefix*	*Symbol*	*Multiple*	*Prefix*	*Symbol*
10^{-1}	deci	d	10	deca	da
10^{-2}	centi	c	10^{2}	hecto	h
10^{-3}	milli	m	10^{3}	kilo	k
10^{-6}	micro	μ	10^{6}	mega	M
10^{-9}	nano	n	10^{9}	giga	G
10^{-12}	pico	p	10^{12}	tera	T
10^{-15}	femto	f	10^{15}	peta	P
10^{-18}	atto	a	10^{18}	exa	E

Prefix symbols should be printed in roman (upright) type with no space between the prefix and the unit symbol.

Example kilometre, km

When a prefix is used with a unit symbol, the combination is taken as a new symbol that can be raised to any power without the use of parentheses.

Examples $1\,\mathrm{cm}^3 = (0.01\ \mathrm{m})^3 = 10^{-6}\ \mathrm{m}^3$
$1\,\mu\mathrm{s}^{-1} = (10^{-6}\ \mathrm{s})^{-1} = 10^{6}\ \mathrm{s}^{-1}$
$1\,\mathrm{V/cm} = 100\ \mathrm{V/m}$
$1\,\mathrm{mmol/dm}^3 = \mathrm{mol\ m}^{-3}$

A prefix should never be used on its own, and prefixes are not to be combined into compound prefixes.

Example pm, not μμm

The names and symbols of decimal multiples and sub-multiples of the SI base unit of mass, the kg, which already contains a prefix, are constructed by adding the appropriate prefix to the word gram and symbol g.

Examples mg, not μkg; Mg, not kkg

The SI prefixes are not to be used with °C.

ISO has recommended standard representations of the prefix symbols for use with limited character sets [6].

3.7 UNITS IN USE TOGETHER WITH THE SI

These units are not part of the SI, but it is recognized that they will continue to be used in appropriate contexts. SI prefixes may be attached to some of these units, such as millilitre, ml; millibar, mbar; megaelectronvolt, MeV; kilotonne, ktonne. A more extensive list of non-SI units, with conversion factors to the corresponding SI units, is given in chapter 7.

Physical quantity	*Name of unit*	*Symbol for unit*	*Value in SI units*
time	minute	min	60 s
time	hour	h	3600 s
time	day	d	86 400 s
plane angle	degree	°	$(\pi/180)$ rad
plane angle	minute	′	$(\pi/10\,800)$ rad
plane angle	second	″	$(\pi/648\,000)$ rad
length	ångström[1]	Å	10^{-10} m
area	barn	b	10^{-28} m^2
volume	litre	l, L	$dm^3 = 10^{-3}$ m^3
mass	tonne	t	Mg $= 10^3$ kg
pressure	bar[1]	bar	10^5 Pa $= 10^5$ N m^{-2}
energy	electronvolt[2]	eV $(= e \times V)$	$\approx 1.60218 \times 10^{-19}$ J
mass	unified atomic mass unit[2,3]	u $(= m_a(^{12}C)/12)$	$\approx 1.66054 \times 10^{-27}$ kg

(1) The ångström and the bar are approved by CIPM [2] for 'temporary use with SI units', until CIPM makes a further recommendation. However, they should not be introduced where they are not used at present.
(2) The values of these units in terms of the corresponding SI units are not exact, since they depend on the values of the physical constants e (for the electronvolt) and N_A (for the unified atomic mass unit), which are determined by experiment. See chapter 5.
(3) The unified atomic mass unit is also sometimes called the dalton, with symbol Da, although the name and symbol have not been approved by CGPM.

3.8 ATOMIC UNITS [8] (see also section 7.3, p.114)

For the purposes of quantum mechanical calculations of electronic wavefunctions, it is convenient to regard certain fundamental constants (and combinations of such constants) as though they were units. They are customarily called *atomic units* (abbreviated: au), and they may be regarded as forming a coherent system of units for the calculation of electronic properties in theoretical chemistry, although there is no authority from CGPM for treating them as units. They are discussed further in relation to the electromagnetic units in chapter 7, p.114. The first five atomic units in the table below have special names and symbols. Only four of these are independent; all others may be derived by multiplication and division in the usual way, and the table includes a number of examples.

The relation of atomic units to the corresponding SI units involves the values of the fundamental physical constants, and is therefore not exact. The numerical values in the table are based on the estimates of the fundamental constants given in chapter 5. The numerical results of calculations in theoretical chemistry are frequently quoted in atomic units, or as numerical values in the form (*physical quantity*)/(*atomic unit*), so that the reader may make the conversion using the current best estimates of the physical constants.

Physical quantity	*Name of unit*	*Symbol for unit*	*Definition and value of unit in SI*
mass	electron rest mass	m_e	$m_e \approx 9.1095 \times 10^{-31}$ kg
charge	elementary charge	e	$e \approx 1.6022 \times 10^{-19}$ C
action	Planck constant/2π	$\hbar$	$\hbar = h/2\pi \approx 1.0546 \times 10^{-34}$ J s
length	bohr	a_0	$4\pi\varepsilon_0\hbar^2/m_e e^2 \approx 5.2918 \times 10^{-11}$ m
energy	hartree	E_h	$\hbar^2/m_e a_0^2 \approx 4.3598 \times 10^{-18}$ J
time	au of time	$\hbar/E_h$	$\approx 2.4189 \times 10^{-17}$ s
velocity[1]	au of velocity	$a_0 E_h/\hbar$	$\approx 2.1877 \times 10^{6}$ m s^{-1}
force	au of force	E_h/a_0	$\approx 8.2389 \times 10^{-8}$ N
momentum, linear	au of momentum	$\hbar/a_0$	$\approx 1.9929 \times 10^{-24}$ N s
electric current	au of current	$eE_h/\hbar$	$\approx 6.6236 \times 10^{-3}$ A
electric field	au of electric field	E_h/ea_0	$\approx 5.1422 \times 10^{11}$ V m^{-1}
electric dipole moment	au of electric dipole	ea_0	$\approx 8.4784 \times 10^{-30}$ C m
magnetic flux density	au of magnetic flux density	$\hbar/ea_0^2$	$\approx 2.3505 \times 10^{5}$ T
magnetic dipole moment[2]	au of magnetic dipole	$e\hbar/m_e$	$= 2\mu_B \approx 1.8548 \times 10^{-23}$ J T^{-1}

(1) The numerical value of the speed of light, when expressed in atomic units, is equal to the reciprocal of the fine structure constant α; c/(au of velocity) $= c\hbar/a_0 E_h = \alpha^{-1} \approx 137.04$.
(2) The atomic unit of magnetic dipole moment is twice the Bohr magneton, μ_B.

4

Recommended mathematical symbols

4.1 PRINTING OF NUMBERS AND MATHEMATICAL SYMBOLS [4.a]

(i) Numbers in general should be printed in roman (upright) type. The decimal sign between digits in a number should be a point (e.g. 2.3) or a comma (e.g. 2,3). To facilitate the reading of long numbers the digits may be grouped in threes about the decimal sign but no point or comma should be used except for the decimal sign. When the decimal sign is placed before the first significant digit of a number a zero should always precede the decimal sign.

Examples 2 573.421 736 or 2 573,421 736 or 0.2573×10^4 or $0{,}2573 \times 10^4$

(ii) Numerical values of physical quantities which have been experimentally determined are usually subject to some uncertainty. The experimental uncertainty should always be specified. The magnitude of the uncertainty may be represented as follows.

Examples $l = 5.3478 \pm 0.0065$ cm
or $l = 5.3478\,(65)$ cm
or $l = 5.34_8$ cm

In the first example, the range of uncertainty (± 0.0065 cm) might be read as the standard deviation (standard error) in the value of l, or more simply as the extreme range of uncertainty to be expected in l; it is desirable to specify the meaning intended.[1]

In the second example the range of uncertainty, indicated in parentheses, is assumed to apply to the least significant digits quoted, so that it is a simpler way of conveying the same information; in both the first two cases the range is ± 0.0065 cm. The third notation implies a less precise estimate of uncertainty, which would be read as between 1 and 9 in the subscripted digit.

(iii) Letter symbols for mathematical constants (e.g. e, π, $\mathrm{i} = \sqrt{-1}$) should be printed in roman (upright) type, but letter symbols for numbers other than constants (e.g. quantum numbers) should be printed in italic (sloping) type, similar to physical quantities.

(iv) Symbols for special mathematical functions (e.g. log, lg, exp, sin, cos, d, δ, Δ, ∇, . . .) should be printed in roman type, but symbols for a general function (e.g. $f(x)$, $F(x, y)$, . . .) should be printed in italic type.

(v) Symbols for symmetry species in group theory (e.g. S, P, D, . . . , s, p, d, . . . , Σ, Π, Δ, . . . , $\mathrm{A_{1g}}$, $\mathrm{B_2''}$, . . .) should be printed in roman (upright) type when they represent the state symbol for an atom or a molecule, although they are often printed in italic type when they represent the symmetry species of a point group.

(vi) Vectors and matrices should be printed in bold face italic type.

Examples force ***F***, electric field ***E***, vector coordinate ***r***

---

(1) The symbol ± is used here with a different meaning to that implied when ± is used as a mathematical operator; the numerical value with its range of uncertainty is to be regarded as specifying a single number. It is not usual to enclose the number 5.3478 ± 0.0065 in parentheses before the unit symbol, although it would not be incorrect to do so.

Ordinary italic type is used to denote the magnitude of the corresponding vector.

Example $r = |\boldsymbol{r}|$.

Tensor quantities may be printed in bold face italic sans-serif type.

Examples $\boldsymbol{\mathsf{S}}$, $\boldsymbol{\mathsf{T}}$.

4.2 SYMBOLS, OPERATORS, AND FUNCTIONS [4.1]

equal to	$=$
not equal to	$\neq$
identically equal to	$\equiv$
equal by definition to	$\stackrel{\text{def}}{=}$
approximately equal to	$\approx$
asymptotically equal to	$\simeq$
corresponds to	$\triangleq$
proportional to	$\propto$, $\sim$
tend to, approaches	$\rightarrow$
infinity	∞
less than	$<$
greater than	$>$
less than or equal to	$\leqslant$
greater than or equal to	$\geqslant$
much less than	$\ll$
much greater than	$\gg$
plus	$+$
minus	$-$
plus or minus	$\pm$
minus or plus	$\mp$

a multiplied by b[1]	ab, $a \cdot b$, $a \times b$
a divided by b	a/b, ab^{-1}, $\frac{a}{b}$
magnitude of a	$\lvert a \rvert$
a to the power n	a^n
square root of a, and of a^2+b^2	$\sqrt{a}$, $a^{1/2}$, $\sqrt{a^2+b^2}$, $(a^2+b^2)^{1/2}$
nth root of a	$a^{1/n}$, $\sqrt[n]{a}$
mean value of a	$\langle a \rangle$, $\bar{a}$
sign of a (equal to $a/\lvert a \rvert$)	$\operatorname{sgn} a$
n factorial	$n!$
binomial coefficient $= n!/p!(n-p)!$	C_p^n, $\binom{n}{p}$
sum of a_i	$\sum a_i$, $\sum_i a_i$, $\sum_{i=1}^{n} a_i$
product of a_i	$\prod a_i$, $\prod_i a_i$, $\prod_{i=1}^{n} a_i$
sine of x	$\sin x$
cosine of x	$\cos x$
tangent of x	$\tan x$
cotangent of x	$\cot x$
inverse sine of x	$\arcsin x$
inverse cosine of x	$\arccos x$
inverse tangent of x	$\arctan x$
hyperbolic sine of x	$\sinh x$
hyperbolic cosine of x	$\cosh x$
hyperbolic tangent of x	$\tanh x$
hyperbolic cotangent of x	$\coth x$
base of natural logarithms	e
exponential of x	$\exp x$, e^x
natural logarithm of x	$\ln x$, $\log_{\mathrm{e}} x$
logarithm to the base a of x	$\log_a x$
logarithm to the base 10 of x	$\lg x$, $\log_{10} x$
logarithm to the base 2 of x	$\operatorname{lb} x$, $\log_2 x$

(1) When multiplication is indicated by a dot, the dot should be raised: $a \cdot b$

square root of minus one	i
real part of $z=a+ib$	Re z
imaginary part of $z=a+ib$	Im z
modulus of $z=a+ib$, absolute value of $z=a+ib$	$\|z\|=(a^2+b^2)^{1/2}$
argument of $z=a+ib$	$\arg z=\arctan(b/a)$
complex conjugate of $z=a+ib$	$z^*=a-ib$
greatest integer $\leqslant x$	ent x, int x
integer division, ent(n/m)	n div m
remainder after integer division, $n/m-$ent(n/m)	n mod m
change in x	$\Delta x=x$ (final) $-x$ (initial)
infinitesimal change of f	δf
limit of $f(x)$ as x tends to a	$\lim_{x\to a} f(x)$
1st derivative of f	$\mathrm{d}f/\mathrm{d}x$, $\partial_x f$, $\mathrm{D}_x f$, f'
nth derivative of f	$\mathrm{d}^n f/\mathrm{d}x^n$, $f'' \cdots$
partial derivative of f	$\partial f/\partial x$
total differential of f	$\mathrm{d}f$
first derivative of x with respect to time	$\dot{x}$, $\partial x/\partial t$
integral of $f(x)$	$\int f(x)\,\mathrm{d}x$
Kronecker delta	$\delta_{ij}=1$ if $i=j$, $=0$ if $i\neq j$
Levi–Civita symbol	$\varepsilon_{ijk}=1$ if i, j, k is a cyclic permutation, $=-1$ if i, j, k is anticyclic, $=0$ otherwise.
Dirac delta function	$\delta(x)$, $\int f(x)\,\delta(x)\,\mathrm{d}x=f(0)$
unit step function	$\varepsilon(x)=1$ for $x>0$, $=0$ for $x<0$
gamma function	$\Gamma(x)=\int t^{x-1}\,\mathrm{e}^{-t}\,\mathrm{d}t$ $=(x-1)!$ for integer values of x
convolution of functions f and g	$f*g=\int f(x-x')g(x')\,\mathrm{d}x'$

vectors

vector a	$\boldsymbol{a}$, $(\vec{a})$
cartesian components of a	a_x, a_y, a_z
unit vectors in cartesian axes	$\boldsymbol{i}, \boldsymbol{j}, \boldsymbol{k}$, or $\boldsymbol{e}_x$, $\boldsymbol{e}_y$, $\boldsymbol{e}_z$
scalar product	$\boldsymbol{a}\cdot\boldsymbol{b}$
vector or cross product	$\boldsymbol{a}\times\boldsymbol{b}$, $\boldsymbol{a}\wedge\boldsymbol{b}$
nabla operator	$\boldsymbol{\nabla}=\boldsymbol{i}\partial/\partial x+\boldsymbol{j}\partial/\partial y+\boldsymbol{k}\partial/\partial z$
Laplacian operator	∇^2, $\Delta=\partial^2/\partial x^2+\partial^2/\partial y^2+\partial^2/\partial z^2$
gradient of a scalar field V	grad V, $\boldsymbol{\nabla} V$
divergence of a vector field A	div $\boldsymbol{A}$, $\boldsymbol{\nabla}\cdot\boldsymbol{A}$
curl of a vector field A	curl $\boldsymbol{A}$, rot $\boldsymbol{A}$, $\boldsymbol{\nabla}\times\boldsymbol{A}$

matrices

matrix of elements A_{ij}	$\boldsymbol{A}$
product of matrices $\boldsymbol{A}$ and $\boldsymbol{B}$	$\boldsymbol{AB}$, $(AB)_{ik}=\sum_j A_{ij}B_{jk}$
(double) scalar product of $\boldsymbol{A}$ and $\boldsymbol{B}$	$\boldsymbol{A}:\boldsymbol{B}=\sum_{i,j} A_{ij}B_{ji}$
unit matrix	$\boldsymbol{E}$, $\boldsymbol{I}$

inverse of a square matrix $\boldsymbol{A}$	$\boldsymbol{A}^{-1}$
transpose of matrix $\boldsymbol{A}$	$\boldsymbol{A}^{\mathrm{T}}$, $\tilde{\boldsymbol{A}}$, $\boldsymbol{A}'$
complex conjugate of matrix $\boldsymbol{A}$	$\boldsymbol{A}^*$
conjugate transpose of $\boldsymbol{A}$ (hermitian conjugate of $\boldsymbol{A}$)	$\boldsymbol{A}^{\dagger}$, $(\boldsymbol{A}^{\dagger})_{ij}=A_{ji}{}^*$
trace of square matrix $\boldsymbol{A}$	$\mathrm{tr}\ A$, $\mathrm{Tr}(A)$, $\Sigma_i\, A_{ii}$
determinant of square matrix $\boldsymbol{A}$	$\det A$, $\lvert A \rvert$

logical operators

A is contained in B	$A \subset B$
union of A and B	$A \cup B$
intersection of A and B	$A \cap B$
p and q (conjunction sign)	$p \wedge q$
p or q or both (disjunction sign)	$p \vee q$
x belongs to A	$x \in A$
x does not belong to A	$x \notin A$
the set A contains x	$A \ni x$
difference of A and B	$A \backslash B$

5
Fundamental physical constants

The following values were recommended by the CODATA Task Group on Fundamental constants in 1986 [51]. For each constant the standard deviation uncertainty in the least significant digits is given in parentheses.

Quantity	*Symbol*	*Value*
permeability of vacuum	μ_0	$4\pi \times 10^{-7}$ H m^{-1} exactly
speed of light in vacuum	c_0	299 792 458 m s^{-1} exactly
permittivity of vacuum[1]	$\varepsilon_0 = 1/\mu_0 c_0{}^2$	8.854 187 816 . . . $\times 10^{-12}$ F m^{-1}
Planck constant	h	6.626 075 5 (40) $\times 10^{-34}$ J s
	$\hbar = h/2\pi$	1.054 572 66 (63) $\times 10^{-34}$ J s
elementary charge	e	1.602 177 33 (49) $\times 10^{-19}$ C
electron rest mass	m_e	9.109 389 7 (54) $\times 10^{-31}$ kg
proton rest mass	m_p	1.672 623 1 (10) $\times 10^{-27}$ kg
neutron rest mass	m_n	1.674 928 6 (10) $\times 10^{-27}$ kg
atomic mass constant (unified atomic mass unit)	$m_u = 1$ u	1.660 540 2 (10) $\times 10^{-27}$ kg
Avogadro constant	L, N_A	6.022 136 7 (36) $\times 10^{23}$ mol^{-1}
Boltzmann constant	k	1.380 658 (12) $\times 10^{-23}$ J K^{-1}
Faraday constant	F	9.648 530 9 (29) $\times 10^{4}$ C mol^{-1}
gas constant	R	8.314 510 (70) J K^{-1} mol^{-1}
zero of the Celsius scale		273.15 K exactly
molar volume, ideal gas, $p = 1$ bar, $\theta = 0\,°C$		22.711 08 (19) L mol^{-1}
standard atmosphere	atm	101 325 Pa exactly
fine structure constant	$\alpha = \mu_0 e^2 c/2h$	7.297 353 08 (33) $\times 10^{-3}$
	α^{-1}	137.035 989 5 (61)
Bohr radius	$a_0 = 4\pi\varepsilon_0 \hbar^2/m_e e^2$	5.291 772 49 (24) $\times 10^{-11}$ m
Hartree energy	$E_h = \hbar^2/m_e a_0{}^2$	4.359 748 2 (26) $\times 10^{-18}$ J
Rydberg constant	$R_\infty = E_h/2hc$	1.097 373 153 4 (13) $\times 10^{7}$ m^{-1}
Bohr magneton	$\mu_B = e\hbar/2m_e$	9.274 015 4 (31) $\times 10^{-24}$ J T^{-1}
electron magnetic moment	μ_e	9.284 770 1 (31) $\times 10^{-24}$ J T^{-1}
Landé g factor for free electron	$g_e = 2\mu_e/\mu_B$	2.002 319 304 386 (20)
nuclear magneton	$\mu_N = (m_e/m_p)\mu_B$	5.050 786 6 (17) $\times 10^{-27}$ J T^{-1}
proton magnetic moment	μ_p	1.410 607 61 (47) $\times 10^{-26}$ J T^{-1}
proton magnetogyric ratio	γ_p	2.675 221 28 (81) $\times 10^{8}$ s^{-1} T^{-1}
magnetic moment of protons in H_2O, μ'_p	μ'_p/μ_B	1.520 993 129 (17) $\times 10^{-3}$
proton resonance frequency per field in H_2O	$\gamma'_p/2\pi$	42.576 375 (13) MHz T^{-1}
Stefan–Boltzmann constant	$\sigma = 2\pi^5 k^4/15h^3c^2$	5.670 51 (19) $\times 10^{-8}$ W m^{-2} K^{-4}
first radiation constant	$c_1 = 2\pi hc^2$	3.741 774 9 (22) $\times 10^{-16}$ W m^2
second radiation constant	$c_2 = hc/k$	1.438 769 (12) $\times 10^{-2}$ m K
gravitational constant	G	6.672 59 (85) $\times 10^{-11}$ m^3 kg^{-1} s^{-2}
standard acceleration of free fall	g_n	9.806 65 m s^{-2} exactly

(1) ε_0 may be calculated exactly from the exact values of μ_0 and c_0.

Quantity	*Symbol*	*Value*
Accurate values of common mathematical constants		
ratio of circumference to diameter of a circle[2]	π	3.141 592 653 59
base of natural logarithms	e	2.718 281 828 46
natural logarithm of 10	ln 10	2.302 585 092 99

(2) A mnemonic for π, based on the numbers of letters in words of the English language, is: 'How I like a drink, alcoholic of course, after the heavy chapters involving quantum mechanics . . .'. A similar mnemonic based on the French language, is: 'Que j'aime a faire apprendre un nombre, utile aux sages . . .'.

6
Properties of particles, elements and nuclides

The symbols for particles, chemical elements and nuclides have been discussed in section 2.10. The recently recommended systematic nomenclature and symbolism for chemical elements of atomic number greater than 103 is briefly described in footnote U to table 6.2.

6.1 PROPERTIES OF SOME PARTICLES

The data given in the table are taken from the compilations by Cohen and Taylor [51], the Particle Data Group [52] and by Wapstra and Audi [53].

Name	*Symbol*	*Spin* I	*Charge number* Z	*Rest mass* m/u	mc^2/MeV	*Magnetic moment* μ/μ_N	*Mean life* τ/s
photon	γ	1	0	$<3\times10^{-36}$	$<3\times10^{-33}$		stable
neutrino	ν_e	1/2	0	$<4.9\times10^{-8}$	$<4.6\times10^{-5}$		stable
electron	e	1/2	-1	$5.485\,799\,03(13)\times10^{-4}$	0.510 999 06 (15)	1.001 159 652 209 (31)[1]	$>6\times10^{29}$
muon	$\mu^\pm$	1/2	±1	0.113 428 913 (17)	105.658 389 (34)	1.001 165 923 (9)[2]	$2.197\,09(5)\times10^{-6}$
pion	$\pi^\pm$	1	±1	0.149 830 4 (9)	139.566 17 (84)		$2.603\,0(23)\times10^{-8}$
pion	π^0	1	0	0.144 830 4 (9)	134.908 70 (84)		$0.83(6)\times10^{-16}$
proton	p	1/2	1	1.007 276 470 (12)	938.272 31 (28)	2.792 844 4 (11)	$>5\times10^{32}$
neutron	n	1/2	0	1.008 664 904 (14)	939.565 63 (28)	$-1.913\,043\,08(54)$	898 (16)
deuteron	d	1	1	2.013 553 214 (24)	1875.613 39 (53)	0.857 437 6 (1)	
triton	t	1/2	1	3.015 500 71 (4)	2808.921 78 (85)	2.978 960 (1)	
helion	h	1/2	2	3.014 932 23 (4)	2808.392 25 (85)	$-2.127\,624(1)$	
α-particle	α	0	2	4.001 506 170 (50)	3727.380 3 (11)	0	

(1) The value is given in Bohr magnetons μ/μ_B, $\mu_B = e\hbar/2m_e$.
(2) The value is given as μ/μ_μ where $\mu_\mu = e\hbar/2m_\mu$.

In nuclear physics and chemistry the masses of particles are often quoted as their energy equivalents (usually in megaelectronvolts). The unified atomic mass unit corresponds to 931.494 32 (28) MeV [51].

The names of pairs of a particle and its antiparticle are derived from the names of the positive particle by adding the ending -ium.

Examples positronium (e^+e^-) $m(e^+e^-) = 1.097\,152\,503(26)\times10^{-3}$ u
muonium ($\mu^+\mu^-$) $m(\mu^+\mu^-) = 0.226\,856\,316(34)$ u

The positive or negative sign for the magnetic moment of a particle implies that the orientation of the magnetic dipole with respect to the angular momentum corresponds to the rotation of a positive or negative charge respectively.

6.2 STANDARD ATOMIC WEIGHTS OF THE ELEMENTS 1985

As agreed by the IUPAC Commission on Atomic Weights and Isotopic Abundances in 1979 [31] the relative atomic mass (atomic weight) of an element, E, can be defined for any specified sample. It is the average mass of its atoms in the sample divided by the unified atomic mass unit[1] or alternatively the molar mass of its atoms divided by the standard molar mass $M^{\circ} = Lm_u = 1\ \text{g mol}^{-1}$:

$$A_r(\text{E}) = \bar{m}_a(\text{E})/\text{u} = M(\text{E})/M^{\circ}.$$

The variations in isotopic composition of many elements in samples of different origin limit the precision to which a relative atomic mass can be given. The standard atomic weights revised biennially by the IUPAC Commission on Atomic Weights and Isotopic Abundances are meant to be applicable for normal materials. This means that to a high level of confidence the relative atomic mass of an element in any normal sample will be within the uncertainty limits of the tabulated value. By 'normal' it is meant here that the material is a reasonably possible source of the element or its compounds in commerce for industry and science and that it has not been subject to significant modification of isotopic composition within a geologically brief period [32]. This, of course, excludes materials studied themselves for very anomalous isotopic composition.

Table 6.2 lists the relative atomic masses of the elements in the alphabetical order of chemical symbols. The values have been recommended by the IUPAC Commission on Atomic Weights and Isotopic Abundances in 1985 [33] and apply to elements as they exist naturally on earth.

The relative atomic masses of many elements depend on the origin and treatment of the materials [34]. The notes to this table explain the types of variation to be expected for individual elements. When used with due regard to the notes the values are considered reliable to ± the figure given in parentheses being applicable to the last digit. For elements without a characteristic terrestrial isotopic composition no standard atomic weight is recommended. The atomic mass of its most stable isotope can be found in table 6.3.

Symbol	*Atomic number*	*Name*	*Relative atomic mass (atomic weight)*	*Note*
Ac	89	actinium		A
Ag	47	silver	107.868 2 (2)	g
Al	13	aluminium	26.981 539 (5)	
Am	95	americium		A
Ar	18	argon	39.948 (1)	g, r
As	33	arsenic	74.921 59 (2)	
At	85	astatine		A
Au	79	gold	196.966 54 (3)	
B	5	boron	10.811 (5)	g, m, r
Ba	56	barium	137.327 (7)	
Be	4	beryllium	9.012 182 (3)	
Bi	83	bismuth	208.980 37 (3)	
Bk	97	berkelium		A
Br	35	bromine	79.904 (1)	
C	6	carbon	12.011 (1)	r

(1) Note that the atomic mass constant, m_u, is equal to the unified atomic mass unit, u, and is defined in terms of the mass of the carbon-12 atom: $m_u = 1\ \text{u} = m_a(^{12}\text{C})/12$.

Symbol	*Atomic number*	*Name*	*Relative atomic mass (atomic weight)*	*Note*
Ca	20	calcium	40.078 (4)	g
Cd	48	cadmium	112.411 (8)	g
Ce	58	cerium	140.115 (4)	g
Cf	98	californium		A
Cl	17	chlorine	35.452 7 (9)	
Cm	96	curium		A
Co	27	cobalt	58.933 20 (1)	
Cr	24	chromium	51.996 1 (6)	
Cs	55	caesium	132.905 43 (5)	
Cu	29	copper	63.546 (3)	r
Dy	66	dysprosium	162.50 (3)	g
Er	68	erbium	167.26 (3)	g
Es	99	einsteinium		A
Eu	63	europium	151.965 (9)	g
F	9	fluorine	18.998 403 2 (9)	
Fe	26	iron	55.847 (3)	
Fm	100	fermium		A
Fr	87	francium		A
Ga	31	gallium	69.723 (4)	
Gd	64	gadolinium	157.25 (3)	g
Ge	32	germanium	72.61 (2)	
H	1	hydrogen	1.007 94 (7)	g, m, r
He	2	helium	4.002 602 (2)	g, r
Hf	72	hafnium	178.49 (2)	
Hg	80	mercury	200.59 (3)	
Ho	67	holmium	164.930 32 (3)	
I	53	iodine	126.904 47 (3)	
In	49	indium	114.82 (1)	
Ir	77	iridium	192.22 (3)	
K	19	potassium	39.098 3 (1)	
Kr	36	krypton	83.80 (1)	g, m
La	57	lanthanum	138.905 5 (2)	g
Li	3	lithium	6.941 (2)	g, m, r
Lr	103	lawrencium		A
Lu	71	lutetium	174.967 (1)	g
Md	101	mendelevium		A
Mg	12	magnesium	24.305 0 (6)	
Mn	25	manganese	54.938 05 (1)	
Mo	42	molybdenum	95.94 (1)	
N	7	nitrogen	14.006 74 (7)	g, r
Na	11	sodium	22.989 768 (6)	
Nb	41	niobium	92.906 38 (2)	
Nd	60	neodymium	144.24 (3)	g
Ne	10	neon	20.179 7 (6)	g, m
Ni	28	nickel	58.69 (1)	

Symbol	Atomic number	Name	Relative atomic mass (atomic weight)	Note
No	102	nobelium		A
Np	93	neptunium		A
O	8	oxygen	15.999 4 (3)	g, r
Os	76	osmium	190.2 (1)	g
P	15	phosphorus	30.973 762 (4)	
Pa	91	protactinium	231.035 88 (2)	Z
Pb	82	lead	207.2 (1)	g, r
Pd	46	palladium	106.42 (1)	g
Pm	61	promethium		A
Po	84	polonium		A
Pr	59	praseodymium	140.907 65 (3)	
Pt	78	platinum	195.08 (3)	
Pu	94	plutonium		A
Ra	88	radium		A
Rb	37	rubidium	85.467 8 (3)	g
Re	75	rhenium	186.207 (1)	
Rh	45	rhodium	102.905 50 (3)	
Rn	86	radon		A
Ru	44	ruthenium	101.07 (2)	g
S	16	sulfur	32.066 (6)	r
Sb	51	antimony	121.75 (3)	
Sc	21	scandium	44.955 910 (9)	
Se	34	selenium	78.96 (3)	
Si	14	silicon	28.085 5 (3)	r
Sm	62	samarium	150.36 (3)	g
Sn	50	tin	118.710 (7)	g
Sr	38	strontium	87.62 (1)	g, r
Ta	73	tantalum	180.947 9 (1)	
Tb	65	terbium	158.925 34 (3)	
Tc	43	technetium		A
Te	52	tellurium	127.60 (3)	g
Th	90	thorium	232.038 1 (1)	g, r, Z
Ti	22	titanium	47.88 (3)	
Tl	81	thallium	204.383 3 (2)	
Tm	69	thulium	168.934 21 (3)	
U	92	uranium	238.028 9 (1)	g, m, Z
Une	109	unnilennium		A, U
Unh	106	unnilhexium		A, U
Uno	108	unniloctium		A, U
Unp	105	unnilpentium		A, U
Unq	104	unnilquadium		A, U
Uns	107	unnilseptium		A, U
V	23	vanadium	50.941 5 (1)	
W	74	tungsten	183.85 (3)	
Xe	54	xenon	131.29 (2)	g, m
Y	39	yttrium	88.905 85 (2)	

Symbol	*Atomic number*	*Name*	*Relative atomic mass (atomic weight)*	*Note*
Yb	70	ytterbium	173.04(3)	g
Zn	30	zinc	65.39(2)	
Zr	40	zirconium	91.224(2)	g

(g) **g**eologically exceptional specimens are known in which the element has an isotopic composition outside the limits for normal material. The difference between the average relative atomic mass of the element in such specimens and that given in the table may exceed considerably the implied uncertainty.
(m) **m**odified isotopic compositions may be found in commercially available material because it has been subjected to an undisclosed or inadvertent isotopic separation. Substantial deviations in relative atomic mass of the element from that given in the table can occur.
(r) **r**ange in isotopic composition of normal terrestrial material prevents a more precise relative atomic mass being given; tabulated $A_r(E)$ value should be applicable to any normal material.
(A) Radioactive element that lacks a characteristic terrestrial isotopic composition.
(Z) An element without stable nuclide(s), exhibiting a range of characteristic terrestrial compositions of long-lived radionuclide(s) such that a meaningful relative atomic mass can be given.
(U) The names and symbols given here are systematic and based on the atomic numbers of the elements as recommended by the IUPAC Commission of the Nomenclature of Inorganic Chemistry [35]. The names are composed of the following roots representing digits of the atomic number:

1 un,	2 bi,	3 tri,	4 quad,	5 pent,
6 hex,	7 sept,	8 oct,	9 enn,	0 nil

The ending -ium is then added to the three roots. The three-letter symbols are derived from the first letters of the corresponding roots.

6.3 PROPERTIES OF NUCLIDES

The table contains the following properties of naturally occurring and some unstable nuclides:

Column

1 *Z* is the atomic number (number of protons) of the nuclide.

2 Symbol of the element.

3 *A* is the mass number of the nuclide. The * sign denotes an unstable nuclide (for elements without naturally occurring isotopes it is the most stable nuclide) and the # sign a nuclide of sufficiently long lifetime to enable the determination of its isotopic abundance.

4 The atomic mass is given in unified atomic mass units, $u = m_a(^{12}C)/12$, together with the standard errors in parentheses and applicable to the last digits quoted. The data were extracted from a more extensive list of *The 1983 Atomic Mass Evaluation* by Wapstra and Audi [53].

5 Isotopic abundances are given as mole fractions, *x*, of the corresponding atoms in percents. They were recommended in 1983 by the IUPAC Commission on Atomic Weights and Isotopic Abundances [34] and are consistent with the standard atomic weights given in table 6.2. The uncertainties given in parentheses are applicable to the last digits quoted and cover the range of probable variations in the materials as well as experimental errors.

6 *I* is the nuclear spin quantum number.

7 Under magnetic moment the maximum *z*-component expectation value of the magnetic dipole moment, μ, in nuclear magnetons is given. The positive or negative sign implies that the orientation of the magnetic dipole with respect to the angular momentum corresponds to the rotation of a positive or negative charge, respectively. The data were extracted from the compilation by Lederer and Shirley [54] and the uncertainty can be taken as ± 1 in the last digit quoted.

8 Under quadrupole moment, the electric quadrupole moment area (see note 8 on p. 20) is given in units of square femtometres, $fm^2 = 10^{-30}$ m^2, although most of the tables quote them in barns (1 barn $= 10^{-28}$ $m^2 = 100$ fm^2). The positive sign implies a prolate nucleus, the negative sign an oblate nucleus. The data were extracted from the compilation by Lederer and Shirley [54], and the uncertainty can be taken as ± 1 in the last digit quoted or ± 10 when the last digit is underlined.

Z	*Sym-bol*	*A*	*Atomic mass,* m_a/u	*Isotopic abundance,* 100 *x*	*Nuclear spin,* *I*	*Magnetic moment,* μ/μ_N	*Quadrupole moment,* Q/fm^2
1	H	1	1.007 825 035 (12)	99.985 (1)	1/2	+2.792 845 6	
	(D)	2	2.014 101 779 (24)	0.015 (1)	1	+0.857 437 6	+0.2875
	(T)	3*	3.016 049 27 (4)		1/2	+2.978 960	
2	He	3	3.016 029 31 (4)	0.000138 (3)	1/2	−2.127 624	
		4	4.002 603 24 (5)	99.999862 (3)	0	0	
3	Li	6	6.015 121 4 (7)	7.5 (2)	1	+0.822 046 7	−0.0645
		7	7.016 003 0 (9)	92.5 (2)	3/2	+3.256 424	−3.66
4	Be	9	9.012 182 2 (4)	100	3/2	−1.177 9	+5.3
5	B	10	10.012 936 9 (3)	19.9 (2)	3	+1.800 65	+8.473
		11	11.009 305 4 (4)	80.1 (2)	3/2	+2.688 637	+4.065

Z	Symbol	A	Atomic mass, m_a/u	Isotopic abundance, 100 x	Nuclear spin, I	Magnetic moment, μ/μ_N	Quadrupole moment, Q/fm^2
6	C	12	12 (by definition)	98.90(3)	0	0	
		13	13.003 354 826(17)	1.10(3)	1/2	+0.702 411	
		14*	14.003 241 982(27)		0	0	
7	N	14	14.003 074 002(26)	99.634(9)	1	+0.403 760 7	+1.56
		15	15.000 108 97(4)	0.366(9)	1/2	−0.283 189 2	
8	O	16	15.994 914 63(5)	99.762(15)	0	0	
		17	16.999 131 2(4)	0.038(3)	5/2	−1.893 80	−2.578
		18	17.999 160 3(9)	0.200(12)	0	0	
9	F	19	18.998 403 22(15)	100	1/2	+2.628 867	
10	Ne	20	19.992 435 6(22)	90.48(3)	0	0	
		21	20.993 842 8(21)	0.27(1)	3/2	−0.661 796	+10.30
		22	21.991 383 1(18)	9.25(3)	0	0	
11	Na	23	22.989 767 7(10)	100	3/2	+2.217 520	+10.2
12	Mg	24	23.985 042 3(8)	78.99(3)	0	0	
		25	24.985 837 4(8)	10.00(1)	5/2	−0.855 46	+22
		26	25.982 593 7(8)	11.01(2)	0	0	
13	Al	27	26.981 538 6(8)	100	5/2	+3.641 504	+14.0
14	Si	28	27.976 927 1(7)	92.23(1)	0	0	
		29	28.976 494 9(7)	4.67(1)	1/2	−0.555 29	
		30	29.973 770 7(7)	3.10(1)	0	0	
15	P	31	30.973 762 0(6)	100	1/2	+1.131 60	
16	S	32	31.972 070 70(25)	95.02(9)	0	0	
		33	32.971 458 43(23)	0.75(1)	3/2	+0.643 821	−6.4
		34	33.967 866 65(22)	4.21(8)	0	0	
		36	35.967 080 62(27)	0.02(1)	0	0	
17	Cl	35	34.968 852 721(69)	75.77(5)	3/2	+0.821 873 6	−8.249
		37	36.965 902 62(11)	24.23(5)	3/2	+0.684 123 0	−6.493
18	Ar	36	35.967 545 52(29)	0.337(3)	0	0	
		38	37.962 732 5(9)	0.063(1)	0	0	
		40	39.962 383 7(14)	99.600(3)	0	0	
19	K	39	38.963 707 4(12)	93.2581(30)	3/2	+0.391 465 8	+4.9
		40	39.963 999 2(12)	0.0117(1)	4	−1.298 099	−6.7
		41	40.961 825 4(12)	6.7302(30)	3/2	+0.214 869 9	+6.0
20	Ca	40	39.962 590 6(13)	96.941(13)	0	0	
		42	41.958 617 6(13)	0.647(3)	0	0	
		43	42.958 766 2(13)	0.135(3)	7/2	−1.317 27	
		44	43.955 480 6(14)	2.086(5)	0	0	
		46	45.953 689(4)	0.004(3)	0	0	
		48	47.952 533(4)	0.187(3)	0	0	

Z	Sym-bol	A	Atomic mass, m_a/u	Isotopic abundance, $100x$	Nuclear spin, I	Magnetic moment, μ/μ_N	Quadrupole moment, Q/fm^2
21	Sc	45	44.955 910 0 (14)	100	7/2	+4.756 483	−22
22	Ti	46	45.952 629 4 (14)	8.0 (1)	0	0	
		47	46.951 764 0 (11)	7.3 (1)	5/2	−0.788 48	+29
		48	47.947 947 3 (11)	73.8 (1)	0	0	
		49	48.947 871 1 (11)	5.5 (1)	7/2	−1.104 17	+24
		50	49.944 792 1 (12)	5.4 (1)	0	0	
23	V	50	49.947 160 9 (17)	0.250 (2)	6	+3.347 45	7
		51	50.943 961 7 (17)	99.750 (2)	7/2	+5.151 4	−5.2
24	Cr	50	49.946 046 4 (17)	4.345 (9)	0	0	
		52	51.940 509 8 (17)	83.789 (12)	0	0	
		53	52.940 651 3 (17)	9.501 (11)	3/2	−0.474 54	+2.2
		54	53.938 882 5 (17)	2.365 (5)	0	0	
25	Mn	55	54.938 047 1 (16)	100	5/2	+3.453 2	+40
26	Fe	54	53.939 612 7 (15)	5.8 (1)	0	0	
		56	55.934 939 3 (16)	91.72 (30)	0	0	
		57	56.935 395 8 (16)	2.2 (1)	1/2	+0.090 622 94	
		58	57.933 277 3 (16)	0.28 (1)	0	0	
27	Co	59	58.933 197 6 (16)	100	7/2	+4.627	+40.4
28	Ni	58	57.935 346 2 (16)	68.27 (1)	0	0	
		60	59.930 788 4 (16)	26.10 (1)	0	0	
		61	60.931 057 9 (16)	1.13 (1)	3/2	−0.750 02	+16.2
		62	61.928 346 1 (16)	3.59 (1)	0	0	
		64	63.927 967 9 (17)	0.91 (1)	0	0	
29	Cu	63	62.929 598 9 (17)	69.17 (2)	3/2	+2.223 3	−20.9
		65	64.927 792 9 (20)	30.83 (2)	3/2	+2.381 7	−19.5
30	Zn	64	63.929 144 8 (19)	48.6 (3)	0	0	
		66	65.926 034 7 (17)	27.9 (2)	0	0	
		67	66.927 129 1 (17)	4.1 (1)	5/2	+0.875 479	+15.0
		68	67.924 845 9 (18)	18.8 (4)	0	0	
		70	69.925 325 (4)	0.6 (1)	0	0	
31	Ga	69	68.925 580 (3)	60.1 (2)	3/2	+2.016 59	+16.8
		71	70.924 700 5 (25)	39.9 (2)	3/2	+2.562 27	+10.6
32	Ge	70	69.924 249 7 (16)	20.5 (5)	0	0	
		72	71.922 078 9 (16)	27.4 (4)	0	0	
		73	72.923 462 6 (16)	7.8 (2)	9/2	−0.879 466 9	−17.3
		74	73.921 177 4 (15)	36.5 (7)	0	0	
		76	75.921 401 6 (17)	7.8 (2)	0	0	
33	As	75	74.921 594 2 (17)	100	3/2	+1.439 47	+29

Z	*Sym-bol*	A	*Atomic mass,* m_a/u	*Isotopic abundance,* $100\,x$	*Nuclear spin,* I	*Magnetic moment,* μ/μ_N	*Quadrupole moment,* Q/fm^2
34	Se	74	73.922 474 6 (16)	0.9 (1)	0	0	
		76	75.919 212 0 (16)	9.0 (2)	0	0	
		77	76.919 912 5 (16)	7.6 (2)	1/2	+0.535 06	
		78	77.917 307 6 (16)	23.6 (6)	0	0	
		80	79.916 519 6 (19)	49.7 (7)	0	0	
		82	81.916 697 8 (23)	9.2 (5)	0	0	
35	Br	79	78.918 336 1 (26)	50.69 (5)	3/2	+2.106 399	+29.3
		81	80.916 289 (6)	49.31 (5)	3/2	+2.270 560	+27
36	Kr	78	77.920 396 (9)	0.35 (2)	0	0	
		80	79.916 380 (9)	2.25 (2)	0	0	
		82	81.913 482 (6)	11.6 (1)	0	0	
		83	82.914 135 (4)	11.5 (1)	9/2	−0.970 669	+27.0
		84	83.911 507 (4)	57.0 (3)	0	0	
		86	85.910 616 (5)	17.3 (2)	0	0	
37	Rb	85	84.911 794 (3)	72.165 (13)	5/2	+1.353 03	+27.4
		87#	86.909 187 (3)	27.835 (13)	3/2	+2.751 24	+13.2
38	Sr	84	83.913 430 (4)	0.56 (1)	0	0	
		86	85.909 267 2 (28)	9.86 (1)	0	0	
		87	86.908 884 1 (28)	7.00 (1)	9/2	−1.092 83	16
		88	87.905 618 8 (28)	82.58 (1)	0	0	
39	Y	89	88.905 849 (3)	100	1/2	−0.137 415 3	
40	Zr	90	89.904 702 6 (26)	51.45 (2)	0	0	
		91	90.905 643 9 (26)	11.22 (2)	5/2	−1.303 62	
		92	91.905 038 6 (26)	17.15 (1)	0	0	
		94	93.906 314 8 (28)	17.38 (2)	0	0	
		96	95.908 275 (4)	2.80 (1)	0	0	
41	Nb	93	92.906 377 2 (27)	100	9/2	+6.170 5	−37
42	Mo	92	91.906 809 (4)	14.84 (4)	0	0	
		94	93.905 085 3 (26)	9.25 (2)	0	0	
		95	94.905 841 1 (22)	15.92 (4)	5/2	−0.914 2	−1.9
		96	95.904 678 5 (22)	16.68 (4)	0	0	
		97	96.906 020 5 (22)	9.55 (2)	5/2	−0.933 5	−10.2
		98	97.905 407 3 (22)	24.13 (6)	0	0	
		100	99.907 477 (6)	9.63 (2)	0	0	
43	Tc	98*	97.907 215 (4)		6		
44	Ru	96	95.907 599 (8)	5.52 (5)	0	0	
		98	97.905 287 (7)	1.88 (5)	0	0	
		99	98.905 938 9 (23)	12.7 (1)	5/2	−0.641 3	+7.7
		100	99.904 219 2 (24)	12.6 (1)	0	0	
		101	100.905 581 9 (24)	17.0 (1)	5/2	−0.718 9	+44
		102	101.904 348 5 (25)	31.6 (2)	0	0	
		104	103.905 424 (6)	18.7 (2)	0	0	

Z	Symbol	A	Atomic mass, m_a/u	Isotopic abundance, 100 x	Nuclear spin, I	Magnetic moment, μ/μ_N	Quadrupole moment, Q/fm^2
45	Rh	103	102.905 500 (4)	100	1/2	−0.088 40	
46	Pd	102	101.905 634 (5)	1.020 (12)	0	0	
		104	103.904 029 (6)	11.14 (8)	0	0	
		105	104.905 079 (6)	22.33 (8)	5/2	−0.642	+80
		106	105.903 478 (6)	27.33 (5)	0	0	
		108	107.903 895 (4)	26.46 (9)	0	0	
		110	109.905 167 (20)	11.72 (9)	0	0	
47	Ag	107	106.905 092 (6)	51.839 (5)	1/2	−0.113 570	
		109	108.904 756 (4)	48.161 (5)	1/2	−0.130 690 5	
48	Cd	106	105.906 461 (7)	1.25 (3)	0	0	
		108	107.904 176 (6)	0.89 (1)	0	0	
		110	109.903 005 (4)	12.49 (9)	0	0	
		111	110.904 182 (3)	12.80 (6)	1/2	−0.594 885 7	
		112	111.902 757 (3)	24.13 (11)	0	0	
		113#	112.904 400 (3)	12.22 (6)	1/2	−0.622 300 5	
		114	113.903 357 (3)	28.73 (21)	0	0	
		116	115.904 755 (4)	7.49 (9)	0	0	
49	In	113	112.904 061 (4)	4.3 (2)	9/2	+5.528 9	+84.6
		115#	114.903 882 (4)	95.7 (2)	9/2	+5.540 8	+86.1
50	Sn	112	111.904 826 (5)	0.97 (1)	0	0	
		114	113.902 784 (4)	0.65 (1)	0	0	
		115	114.903 348 (3)	0.36 (1)	1/2	−0.918 84	
		116	115.901 747 (3)	14.53 (11)	0	0	
		117	116.902 956 (3)	7.68 (7)	1/2	−1.001 05	
		118	117.901 609 (3)	24.22 (11)	0	0	
		119	118.903 311 (3)	8.58 (4)	1/2	−1.047 29	
		120	119.902 199 1 (29)	32.59 (10)	0	0	
		122	121.903 440 4 (30)	4.63 (3)	0	0	
		124	123.905 274 3 (17)	5.79 (5)	0	0	
51	Sb	121	120.903 821 2 (29)	57.3 (9)	5/2	+3.363 4	−20
		123	122.904 216 0 (24)	42.7 (9)	7/2	+2.549 8	−26
52	Te	120	119.904 048 (21)	0.096 (2)	0	0	
		122	121.903 050 (3)	2.60 (1)	0	0	
		123	122.904 271 0 (22)	0.908 (3)	1/2	−0.736 79	
		124	123.902 818 0 (18)	4.816 (8)	0	0	
		125	124.904 428 5 (25)	7.14 (1)	1/2	−0.888 28	
		126	125.903 309 5 (25)	18.95 (1)	0	0	
		128	127.904 463 (4)	31.69 (2)	0	0	
		130	129.906 229 (5)	33.80 (2)	0	0	
53	I	127	126.904 473 (5)	100	5/2	+2.813 28	−78.9

Z	Symbol	A	Atomic mass, m_a/u	Isotopic abundance, 100 x	Nuclear spin, I	Magnetic moment, μ/μ_N	Quadrupole moment, Q/fm^2
54	Xe	124	123.905 894 2 (22)	0.10 (1)	0	0	
		126	125.904 281 (8)	0.09 (1)	0	0	
		128	127.903 531 2 (17)	1.91 (3)	0	0	
		129	128.904 780 1 (21)	26.4 (6)	1/2	−0.777 977	
		130	129.903 509 4 (17)	4.1 (1)	0	0	
		131	130.905 072 (5)	21.2 (4)	3/2	+0.691 861	−12.0
		132	131.904 144 (5)	26.9 (5)	0	0	
		134	133.905 395 (8)	10.4 (2)	0	0	
		136	135.907 214 (8)	8.9 (1)	0	0	
55	Cs	133	132.905 429 (7)	100	7/2	+2.582 024	−0.3
56	Ba	130	129.906 282 (8)	0.106 (2)	0	0	
		132	131.905 042 (9)	0.101 (2)	0	0	
		134	133.904 486 (7)	2.417 (27)	0	0	
		135	134.905 665 (7)	6.592 (18)	3/2	+0.837 943	+18
		136	135.904 553 (7)	7.854 (39)	0	0	
		137	136.905 812 (6)	11.23 (4)	3/2	+0.937 365	+28
		138	137.905 232 (6)	71.70 (7)	0	0	
57	La	138#	137.907 105 (6)	0.09 (1)	5	+3.713 9	+52
		139	138.906 347 (5)	99.91 (1)	7/2	+2.783 2	+22
58	Ce	136	135.907 140 (50)	0.19 (1)	0	0	
		138	137.905 985 (12)	0.25 (1)	0	0	
		140	139.905 433 (4)	88.48 (10)	0	0	
		142	141.909 241 (4)	11.08 (10)	0	0	
59	Pr	141	140.907 647 (4)	100	5/2	+4.136	−5.89
60	Nd	142	141.907 719 (4)	27.13 (10)	0	0	
		143	142.909 810 (4)	12.18 (5)	7/2	−1.065	−48.4
		144	143.910 083 (4)	23.80 (10)	0	0	
		145	144.912 570 (4)	8.30 (5)	7/2	−0.656	−25.3
		146	145.913 113 (4)	17.19 (8)	0	0	
		148	147.916 889 (4)	5.76 (3)	0	0	
		150	149.920 887 (4)	5.64 (3)	0	0	
61	Pm	145*	144.912 743 (4)		5/2		
62	Sm	144	143.911 998 (4)	3.1 (1)	0	0	
		147#	146.914 894 (4)	15.0 (2)	7/2	−0.814 9	−18
		148	147.914 819 (4)	11.3 (1)	0	0	
		149	148.917 180 (4)	13.8 (1)	7/2	−0.671 8	+5.3
		150	149.917 273 (4)	7.4 (1)	0	0	
		152	151.919 728 (4)	26.7 (2)	0	0	
		154	153.922 205 (4)	22.7 (2)	0	0	
63	Eu	151	150.919 702 (8)	47.8 (5)	5/2	+3.471 8	+116
		153	152.921 225 (4)	52.2 (5)	5/2	+1.533 1	+294

Z	Symbol	A	Atomic mass, m_a/u	Isotopic abundance, 100 x	Nuclear spin, I	Magnetic moment, μ/μ_N	Quadrupole moment, Q/fm^2
64	Gd	152	151.919 786 (4)	0.20 (1)	0	0	
		154	153.920 861 (4)	2.18 (3)	0	0	
		155	154.922 618 (4)	14.80 (5)	3/2	−0.259 1	+159
		156	155.922 118 (4)	20.47 (4)	0	0	
		157	156.923 956 (4)	15.65 (3)	3/2	−0.339 9	+203
		158	157.924 019 (4)	24.84 (12)	0	0	
		160	159.927 049 (4)	21.86 (4)	0	0	
65	Tb	159	158.925 342 (4)	100	3/2	+2.014	118
66	Dy	156	155.924 277 (8)	0.06 (1)	0	0	
		158	157.924 403 (5)	0.10 (1)	0	0	
		160	159.925 193 (4)	2.34 (5)	0	0	
		161	160.926 930 (4)	18.9 (1)	5/2	−0.480 6	+244
		162	161.926 795 (4)	25.5 (2)	0	0	
		163	162.928 728 (4)	24.9 (2)	5/2	+0.672 6	+257
		164	163.929 171 (4)	28.2 (2)	0	0	
67	Ho	165	164.930 319 (4)	100	7/2	+4.173	+274
68	Er	162	161.928 775 (4)	0.14 (1)	0	0	
		164	163.929 198 (4)	1.61 (1)	0	0	
		166	165.930 290 (4)	33.6 (2)	0	0	
		167	166.932 046 (4)	22.95 (13)	7/2	−0.566 5	+282.7
		168	167.932 368 (4)	26.8 (2)	0	0	
		170	169.935 461 (4)	14.9 (1)	0	0	
69	Tm	169	168.934 212 (4)	100	1/2	−0.231 6	
70	Yb	168	167.933 894 (5)	0.13 (1)			
		170	169.934 759 (4)	3.05 (5)	0	0	
		171	170.936 323 (3)	14.3 (2)	1/2	+0.493 67	
		172	171.936 378 (3)	21.9 (3)	0	0	
		173	172.938 208 (3)	16.12 (18)	5/2	−0.679 89	+280
		174	173.938 859 (3)	31.8 (4)	0	0	
		176	175.942 564 (4)	12.7 (1)	0	0	
71	Lu	175	174.940 770 (3)	97.41 (2)	7/2	+2.232 7	+569
		176$^{\#}$	175.942 679 (3)	2.59 (2)	7	+3.19	+810
72	Hf	174	173.940 044 (4)	0.162 (2)	0	0	
		176	175.941 406 (4)	5.206 (4)	0	0	
		177	176.943 217 (3)	18.606 (3)	7/2	+0.793 6	+450
		178	177.943 696 (3)	27.297 (3)	0	0	
		179	178.945 812 2 (29)	13.629 (5)	9/2	−0.640 9	+510
		180	179.946 545 7 (30)	35.100 (6)	0	0	
73	Ta	180	179.947 462 (4)	0.012 (2)	8		
		181	180.947 992 (3)	99.988 (2)	7/2	+2.371	+390

Z	Symbol	A	Atomic mass, m_a/u	Isotopic abundance, 100 x	Nuclear spin, I	Magnetic moment, μ/μ_N	Quadrupole moment, Q/fm^2
74	W	180	179.946 701 (5)	0.13 (3)	0	0	
		182	181.948 202 (3)	26.3 (2)	0	0	
		183	182.950 220 (3)	14.3 (1)	1/2	+0.117 784 7	
		184	183.950 928 (3)	30.67 (15)	0	0	
		186	185.954 357 (4)	28.6 (2)	0	0	
75	Re	185	184.952 951 (3)	37.40 (2)	5/2	+3.187 1	236
		187#	186.955 744 (3)	62.60 (2)	5/2	+3.219 7	224
76	Os	184	183.952 488 (4)	0.02 (1)	0	0	
		186	185.953 830 (4)	1.58 (10)	0	0	
		187	186.955 741 (3)	1.6 (1)	1/2	+0.064 651 85	
		188	187.955 830 (3)	13.3 (2)	0	0	
		189	188.958 137 (4)	16.1 (3)	3/2	+0.659 933	+91
		190	189.958 436 (4)	26.4 (4)	0	0	
		192	191.961 467 (4)	41.0 (3)	0	0	
77	Ir	191	190.960 584 (4)	37.3 (5)	3/2	+0.146 2	78
		193	192.962 917 (4)	62.7 (5)	3/2	+0.159 2	70
78	Pt	190	189.959 917 (7)	0.01 (1)	0	0	
		192	191.961 019 (5)	0.79 (5)	0	0	
		194	193.962 655 (4)	32.9 (5)	0	0	
		195	194.964 766 (4)	33.8 (5)	1/2	+0.609 50	
		196	195.964 926 (4)	25.3 (5)	0	0	
		198	197.967 869 (6)	7.2 (2)	0	0	
79	Au	197	196.966 543 (4)	100	3/2	+0.148 159	54.7
80	Hg	196	195.965 807 (5)	0.14 (10)	0	0	
		198	197.966 743 (4)	10.02 (7)	0	0	
		199	198.968 254 (4)	16.84 (11)	1/2	+0.505 885 2	
		200	199.968 300 (4)	23.13 (11)	0	0	
		201	200.970 277 (4)	13.22 (11)	3/2	−0.560 225	+45.5
		202	201.970 617 (4)	29.80 (14)	0	0	
		204	203.973 467 (5)	6.85 (5)	0	0	
81	Tl	203	202.972 320 (5)	29.524 (9)	1/2	+1.622 257	
		205	204.974 401 (5)	70.476 (9)	1/2	+1.638 213 5	
82	Pb	204	203.973 020 (5)	1.4 (1)	0	0	
		206	205.974 440 (4)	24.1 (1)	0	0	
		207	206.975 872 (4)	22.1 (1)	1/2	+0.582 19	
		208	207.976 627 (4)	52.4 (1)	0	0	
83	Bi	209	208.980 374 (5)	100	9/2	+4.110 6	−46
84	Po	209*	208.982 404 (5)		1/2		
85	At	210*	209.987 126 (12)				
86	Rn	222*	222.017 571 (3)		0	0	

Z	*Symbol*	A	*Atomic mass,* m_a/u	*Isotopic abundance,* 100 x	*Nuclear spin,* I	*Magnetic moment,* μ/μ_N	*Quadrupole moment,* Q/fm^2
87	Fr	223*	223.019 733 (4)		3/2		
88	Ra	226*	226.025 403 (3)		0	0	
89	Ac	227*	227.027 750 (3)		3/2	+1.1	$+17\underline{0}$
90	Th	$232^{\#}$	232.038 050 8 (23)	100	0	0	
91	Pa	231*	231.035 880 (3)		3/2	2.01	
92	U	$234^{\#}$	234.040 946 8 (24)	0.0055 (5)	0	0	
		$235^{\#}$	235.043 924 2 (24)	0.7200 (12)	7/2	−0.35	455
		$238^{\#}$	238.050 784 7 (23)	99.274 5 (15)	0	0	
93	Np	237*	237.048 167 8 (23)		5/2	+3.14	$+42\underline{0}$
94	Pu	244*	244.064 199 (5)		0		
95	Am	243*	243.061 375 (3)		5/2	+1.61	$+49\underline{0}$
96	Cm	247*	247.070 347 (5)				
97	Bk	247*	247.070 300 (6)				
98	Cf	251*	251.079 580 (5)				
99	Es	252*	252.082 944 (23)				
100	Fm	257*	257.095 099 (8)				
101	Md	258*	258.098 57 (22)				
102	No	259*	259.100 931 (12)				
103	Lr	260*	260.105 320 (60)				
104	Unq	261*	261.108 69 (22)				
105	Unp	262*	262.113 76 (16)				
106	Unh	263*	263.118 22 (13)				
107	Uns	262*	262.122 93 (45)				
108	Uno	265*	265.130 16 (99)				
109	Une	266*	266.137 64 (45)				

7
Conversion of units

SI units are recommended for use throughout science and technology. However, some non-SI units are in use, and in a few cases they are likely to remain so for many years. Moreover the published literature of science makes widespread use of non-SI units. It is thus often necessary to convert the values of physical quantities between SI and other units. This chapter is concerned with facilitating this process.

Section 7.1 gives examples illustrating the use of quantity calculus for converting the values of physical quantities between different units. The table in section 7.2 lists a variety of non-SI units used in chemistry, with the conversion factors to the corresponding SI units. Conversion factors for energy and energy-related units (wavenumber, frequency, temperature and molar energy), and for pressure units, are also presented in tables inside the back cover.

Many of the difficulties in converting units between different systems are associated either with the electromagnetic units, or with atomic units and their relationship to the electromagnetic units. In sections 7.3 and 7.4 the relations involving electromagnetic and atomic units are developed in greater detail, to provide a background for the conversion factors presented in the table in section 7.2.

7.1 THE USE OF QUANTITY CALCULUS

Quantity calculus is the system in which we always take the values of physical quantities to be the product of a numerical value and a unit (see section 1.1), and we manipulate the symbols for physical quantities, numerical values, and units by the ordinary rules of algebra.[1] This system is recommended for general use in science. Quantity calculus has particular advantages in facilitating the problems of converting between different units and different systems of units, as illustrated by the examples below. In all of these examples the numerical values are approximate.

Example 1. The wavelength λ of one of the yellow lines of sodium is given by

$$\lambda = 5.896 \times 10^{-7}\ \mathrm{m}, \qquad \text{or} \quad \lambda/\mathrm{m} = 5.896 \times 10^{-7}.$$

The ångström is defined by the equation (see table 7.2, under length):

$$1\,\text{Å} = \text{Å} = 10^{-10}\ \mathrm{m}, \qquad \text{or} \quad \mathrm{m}/\text{Å} = 10^{10}.$$

Substituting in the first equation gives the value of λ in ångström units:

$$\lambda/\text{Å} = (\lambda/\mathrm{m})\ (\mathrm{m}/\text{Å}) = (5.896 \times 10^{-7})\ (10^{10}) = 5896,$$

or

$$\lambda = 5896\ \text{Å}.$$

Example 2. The vapour pressure of water at 20 °C is recorded to be

$$p(\mathrm{H_2O},\ 20\,^{\circ}\mathrm{C}) = 17.5\ \mathrm{Torr}.$$

The Torr, the bar, and the atmosphere are given by the equations (see table 7.2, under pressure)

$$\mathrm{Torr} \approx 133.3\ \mathrm{Pa},$$
$$\mathrm{bar} = 10^5\ \mathrm{Pa},\ \text{and}$$
$$\mathrm{atm} = 101\,325\ \mathrm{Pa}$$

Thus

$$\begin{aligned} p(\mathrm{H_2O},\ 20\,^{\circ}\mathrm{C}) &= 17.5 \times 133.3\ \mathrm{Pa} = 2.33\ \mathrm{kPa} \\ &= (2.33 \times 10^3/10^5)\ \mathrm{bar} = 23.3\ \mathrm{mbar} \\ &= (2.33 \times 10^3/101\,325)\ \mathrm{atm} = 2.30 \times 10^{-2}\ \mathrm{atm}. \end{aligned}$$

Example 3. Spectroscopic measurements show that for the methylene radical, CH_2, the $\tilde{a}\ ^1A_1$ excited state lies at a wavenumber 3156 $\mathrm{cm^{-1}}$ above the $\tilde{X}\ ^3B_1$ ground state,

$$\tilde{\nu}(\tilde{a} - \tilde{X}) = T_0(\tilde{a}) - T_0(\tilde{X}) = 3156\ \mathrm{cm^{-1}}.$$

The excitation energy from the ground triplet state to the excited singlet state is thus

$$\begin{aligned} \Delta E = hc\tilde{\nu} &= (6.626 \times 10^{-34}\ \mathrm{J\,s})\ (2.998 \times 10^8\ \mathrm{m\,s^{-1}})\ (3156\ \mathrm{cm^{-1}}) \\ &= 6.269 \times 10^{-22}\ \mathrm{J\,m\,cm^{-1}} \\ &= 6.269 \times 10^{-20}\ \mathrm{J} = 6.269 \times 10^{-2}\ \mathrm{aJ}, \end{aligned}$$

where the values of h and c are taken from the fundamental physical constants in chapter 5, and we

---

(1) A more appropriate name for 'quantity calculus' might be 'algebra of quantities', because it is the principles of algebra rather than calculus that are involved.

have used the relation $\mathrm{m} = 100\ \mathrm{cm}$, or $\mathrm{m\ cm^{-1}} = 100$. Since the electronvolt is given by the equation (table 7.2, under energy) $\mathrm{eV} \approx 1.6022 \times 10^{-19}\ \mathrm{J}$, or $\mathrm{aJ} \approx (1/0.16022)\ \mathrm{eV}$,

$$\Delta E = (6.269 \times 10^{-2}/0.16022)\ \mathrm{eV} = 0.3913\ \mathrm{eV}.$$

Similarly the Hartree energy is given by (table 7.3) $E_\mathrm{h} = \hbar^2/m_\mathrm{e}a_0{}^2 \approx 4.3598\ \mathrm{aJ}$, or $\mathrm{aJ} \approx (1/4.3598)\ E_\mathrm{h}$, and thus the excitation energy is given in atomic units by

$$\Delta E = (6.269 \times 10^{-2}/4.3598)\ E_\mathrm{h} = 1.4380 \times 10^{-2}\ E_\mathrm{h}.$$

Finally the molar excitation energy is given by

$$\begin{aligned}\Delta E_\mathrm{m} &= N_\mathrm{A}\Delta E \\ &= (6.022 \times 10^{23}\ \mathrm{mol^{-1}})\ (6.269 \times 10^{-2}\ \mathrm{aJ}) \\ &= 37.75\ \mathrm{kJ\ mol^{-1}}\end{aligned}$$

Also, since $\mathrm{kcal} = 4.184\ \mathrm{kJ}$, or $\mathrm{kJ} = (1/4.184)\ \mathrm{kcal}$,

$$\Delta E_\mathrm{m} = (37.75/4.184)\ \mathrm{kcal\ mol^{-1}} = 9.023\ \mathrm{kcal\ mol^{-1}}.$$

Note that in this example the conversion factors are not pure numbers, but have dimensions, and involve the fundamental physical constants h, c, e, m_e, a_0 and N_A. Also in this example the necessary conversion factors could have been taken directly from the table on the inside back cover.

Example 4. The molar conductivity, Λ, of an electrolyte is defined by the equation (see p.53)

$$\Lambda = \kappa/c,$$

where κ is the conductivity of the electrolyte solution minus the conductivity of the pure solvent and c is the electrolyte concentration. Conductivities of electrolytes are usually expressed in $\mathrm{S\ cm^{-1}}$ and concentrations in $\mathrm{mol\ dm^{-3}}$; for example $\kappa(\mathrm{KCl}) = 7.39 \times 10^{-5}\ \mathrm{S\ cm^{-1}}$ for $c(\mathrm{KCl}) = 0.000\,500\ \mathrm{mol\ dm^{-3}}$. The molar conductivity can then be calculated as follows

$$\begin{aligned}\Lambda &= (7.39 \times 10^{-5}\ \mathrm{S\ cm^{-1}})/(0.000\,500\ \mathrm{mol\ dm^{-3}}) \\ &= 0.1478\ \mathrm{S\ mol^{-1}\ cm^{-1}\ dm^3} = 147.8\ \mathrm{S\ mol^{-1}\ cm^2}\end{aligned}$$

since $\mathrm{dm^3} = 1000\ \mathrm{cm^3}$. The above relationship has previously often been, and sometimes still is, written in the form

$$\Lambda = 1000\kappa/c$$

However, in this form the symbols *do not* represent physical quantities, but the *numerical values* of physical quantities in certain units. Specifically, the last equation is true only if Λ is the molar conductivity in $\mathrm{S\ mol^{-1}\ cm^2}$, κ is the conductivity in $\mathrm{S\ cm^{-1}}$, and c is the concentration in $\mathrm{mol\ dm^{-3}}$. This form does not follow the rules of quantity calculus, and should be avoided. The equation $\Lambda = \kappa/c$, in which the symbols represent physical quantities, is true in any units. If it is desired to write the relationship between numerical values it should be written in the form

$$\Lambda/(\mathrm{S\ mol^{-1}\ cm^2}) = \frac{1000\ \kappa/(\mathrm{S\ cm^{-1}})}{c/(\mathrm{mol\ dm^{-3}})}.$$

Example 5. A solution of 0.125 mol of solute B in 953 g of solvent S has a molality m_B given by[2]

$$m_\mathrm{B} = n_\mathrm{B}/m_\mathrm{S} = (0.125/953)\ \mathrm{mol\ g^{-1}} = 0.131\ \mathrm{mol\ kg^{-1}}.$$

---

(2) Note the confusion of notation: m_B denotes molality, and m_S denotes mass. However these symbols are almost always used. See footnote (13) on p.38.

The mole fraction of solute is approximately given by

$$x_B = n_B/(n_S + n_B) \approx n_B/n_S = m_B M_S,$$

where it is assumed that $n_B \ll n_S$.

If the solvent is water with molar mass 18.015 g mol^{-1}, then

$$x_B \approx (0.131\ \text{mol kg}^{-1})\ (18.015\ \text{g mol}^{-1}) = 2.36\ \text{g/kg} = 0.00236.$$

The equations used here are sometimes quoted in the form $m_B = 1000 n_B/m_S$, and $x_B \approx m_B M_S/1000$. However, this is *not* a correct use of quantity calculus because in this form the symbols denote the *numerical values* of the physical quantities in particular units; specifically it is assumed that m_B, m_S and M_S denote numerical values in mol kg^{-1}, g, and g mol^{-1} respectively. A correct way of writing the second equation would, for example, be

$$x_B = (m_B/\text{mol kg}^{-1})(M_S/\text{g mol}^{-1})/1000.$$

Example 6. For paramagnetic materials the magnetic susceptibility may be measured experimentally and used to give information on the molecular magnetic dipole moment, and hence on the electronic structure of the molecules in the material. The paramagnetic contribution to the molar magnetic susceptibility of a material, χ_m, is related to the molecular magnetic dipole moment m by the Curie relation

$$\chi_m = \chi V_m = \mu_0 N_A m^2/3kT.$$

In terms of the irrational susceptibility $\chi^{(ir)}$, which is often used in connection with the older esu, emu, and Gaussian unit systems (see section 7.3 below), this equation becomes

$$\chi_m^{(ir)} = \chi^{(ir)} V_m = (\mu_0/4\pi) N_A m^2/3kT.$$

Solving for m, and expressing the result in terms of the Bohr magneton μ_B,

$$m/\mu_B = (3k/\mu_0 N_A)^{\frac{1}{2}} \mu_B^{-1} (\chi_m T)^{\frac{1}{2}}.$$

Finally, using the values of the fundamental constants μ_B, k, μ_0, and N_A given in chapter 5, we obtain

$$\begin{aligned} m/\mu_B &= 0.7977[\chi_m/(\text{cm}^3\ \text{mol}^{-1})]^{\frac{1}{2}}[T/\text{K}]^{\frac{1}{2}} \\ &= 2.828[\chi_m^{(ir)}/(\text{cm}^3\ \text{mol}^{-1})]^{\frac{1}{2}}[T/\text{K}]^{\frac{1}{2}}. \end{aligned}$$

These expressions are convenient for practical calculations. The final result has frequently been expressed in the form

$$m/\mu_B = 2.828(\chi_m T)^{\frac{1}{2}},$$

where it is assumed, contrary to the conventions of quantity calculus, that χ_m and T denote the *numerical values* of the molar susceptibility and the temperature in the units cm^3 mol^{-1} and K respectively, and where it is also assumed (but rarely stated) that the susceptibility is defined using the irrational electromagnetic equations (see section 7.3 below).

7.2 CONVERSION TABLES FOR UNITS

The table below gives conversion factors from a variety of units to the corresponding SI unit. Examples of the use of this table have already been given in the preceding section. For each physical quantity the name is given, followed by the recommended symbol(s). Then the SI unit is given, followed by the esu, emu, Gaussian unit (Gau), atomic unit (au), and other units in common use, with their conversion factors to SI. The constant ζ which occurs in some of the electromagentic conversion factors is the (exact) pure number $2.997\,924\,58 \times 10^{10} = c_0/(\text{cm s}^{-1})$.

The inclusion of non-SI units in this table should not be taken to imply that their use is to be encouraged. With some exceptions, SI units are always to be preferred to non-SI units. However, since many of the units below are to be found in the scientific literature, it is convenient to tabulate their relation to the SI.

For convenience units in the esu and Gaussian systems are quoted in terms of the four dimensions *length, mass, time,* and *electric charge,* by including the franklin (Fr) as an abbreviation for the electrostatic unit of charge and $4\pi\varepsilon_0$ as a constant with dimensions $(charge)^2/(energy \times length)$. This gives each physical quantity the same dimensions in all systems, so that all conversion factors are pure numbers. The factors $4\pi\varepsilon_0$ and the Fr may be eliminated by writing $\text{Fr} = \text{esu of charge} = \text{erg}^{1/2}\,\text{cm}^{1/2} = \text{cm}^{3/2}\,\text{g}^{1/2}\,\text{s}^{-1}$, and $4\pi\varepsilon_0 = \varepsilon_0^{(\text{ir})} = 1\ \text{Fr}^2\,\text{erg}^{-1}\,\text{cm}^{-1} = 1$, to recover esu expressions in terms of three base units (see section 7.3 below). The symbol Fr should be regarded as a compact representation of (esu of charge).

Conversion factors are either given exactly (when the = sign is used), or they are given to the approximation that the corresponding physical constants are known (when the $\approx$ sign is used). In the latter case the uncertainty is always less than ± 5 in the last digit quoted.

Name	*Symbol*	*Relation to SI*
length, l		
metre (SI unit)	m	
cm (esu, emu, Gau)	cm	$= 10^{-2}$ m
bohr (au)	a_0, b	$= 4\pi\varepsilon_0\hbar^2/m_e e^2 \approx 5.291\,77 \times 10^{-11}$ m
ångström	Å	$= 10^{-10}$ m
micron	μ	$= \mu\text{m} = 10^{-6}$ m
millimicron	mμ	$= \text{nm} = 10^{-9}$ m
x unit	X	$\approx 1.002 \times 10^{-13}$ m
fermi	f	$= \text{fm} = 10^{-15}$ m
inch	in	$= 2.54 \times 10^{-2}$ m
foot	ft	$= 12$ in $= 0.3048$ m
yard	yd	$= 3$ ft $= 0.9144$ m
mile	mi	$= 1760$ yd $= 1609.344$ m
astronomical unit	AU	$= 1.496\,00 \times 10^{11}$ m
parsec	pc	$= 3.085\,68 \times 10^{16}$ m
light year	l.y.	$= 9.460\,53 \times 10^{15}$ m
area, A		
square metre (SI unit)	m^2	
barn	b	$= 10^{-28}\,\text{m}^2$
acre		$= 4046.856\,\text{m}^2$
are	a	$= 100\,\text{m}^2$
hectare	ha	$= 10^4\,\text{m}^2$

Name	*Symbol*	*Relation to SI*
volume, V		
cubic metre (SI unit)	m^3	
litre	l, L	$=dm^3=10^{-3}\ m^3$
lambda	λ	$=\mu l=10^{-6}\ dm^3$
barrel (US)		$=158.987\ dm^3$
gallon (US)	gall (US)	$=3.785\,41\ dm^3$
gallon (UK)	gal (UK)	$=4.546\,09\ dm^3$
mass, m		
kilogram (SI unit)	kg	
gram (esu, emu, Gau)	g	$=10^{-3}\ kg$
electron mass (au)	m_e	$\approx 9.109\,39\times 10^{-31}\ kg$
unified atomic mass unit, dalton	u, Da	$=m_a(^{12}C)/12\approx 1.660\,54\times 10^{-27}\ kg$
gamma	γ	$=\mu g$
tonne	t	$=Mg=10^3\ kg$
pound (avoirdupois)	lb	$=0.453\,592\,37\ kg$
ounce (avoirdupois)	oz	$\approx 28.3495\ g$
ounce (troy)	oz (troy)	$\approx 31.1035\ g$
grain	gr	$=64.798\,91\ mg$
time, t		
second (SI, esu, emu, Gau)	s	
au of time	$\hbar/E_h$	$\approx 2.418\,88\times 10^{-17}\ s$
minute	min	$=60\ s$
hour	h	$=3600\ s$
day	d	$=86\,400\ s$
svedberg	Sv	$=10^{-13}\ s$
acceleration, a		
SI unit	$m\ s^{-2}$	
standard acceleration of free fall	g_n	$=9.806\,65\ m\ s^{-2}$
gal, galileo	Gal	$=10^{-2}\ m\ s^{-2}$
force, F		
newton (SI unit)[1]	N	$=kg\ m\ s^{-2}$
dyne (esu, emu, Gau)	dyn	$=g\ cm\ s^{-2}=10^{-5}\ N$
au of force	E_h/a_0	$\approx 8.238\,73\times 10^{-8}\ N$
kilogram-force	kgf	$=9.806\,65\ N$
energy, U		
joule (SI unit)	J	$=kg\ m^2\ s^{-2}$
erg (esu, emu, Gau)	erg	$=g\ cm^2\ s^{-2}=10^{-7}\ J$
hartree (au)	E_h	$=\hbar^2/m_e a_0^2\approx 4.359\,75\times 10^{-18}\ J$
rydberg	Ry	$=E_h/2\approx 2.179\,87\times 10^{-18}\ J$

(1) 1 N is approximately the force exerted by the earth upon an apple.

Name	*Symbol*	*Relation to SI*
energy, U (continued)		
electronvolt	eV	$= e \times V \approx 1.602\,18 \times 10^{-19}$ J
calorie, thermochemical	cal_{th}	$=4.184$ J
calorie, international	cal_{IT}	$=4.1868$ J
15 °C calorie	cal_{15}	≈ 4.1855 J
litre atmosphere	l atm	$=101.325$ J
British thermal unit	Btu	$=1055.06$ J
pressure, p		
pascal (SI unit)	Pa	$= \mathrm{N\,m^{-2}} = \mathrm{kg\,m^{-1}\,s^{-2}}$
atmosphere	atm	$=101325$ Pa
bar	bar	$=10^5$ Pa
torr	Torr	$=(101325/760)$ Pa ≈ 133.322 Pa
millimetre of mercury (conventional)	mmHg	$=13.5951 \times 980.665 \times 10^{-2}$ Pa ≈ 133.322 Pa
pounds per square inch	psi	$\approx 6.894\,757 \times 10^3$ Pa
power, P		
SI unit	W	$= \mathrm{kg\,m^2\,s^{-3}}$
horse power	hp	$=745.7$ W
action, L, J (angular momentum)		
SI unit	J s	$= \mathrm{kg\,m^2\,s^{-1}}$
esu, emu, Gau	erg s	$=10^{-7}$ J s
au of action	$\hbar$	$= h/2\pi \approx 1.054\,57 \times 10^{-34}$ J s
dynamic viscosity, η		
SI unit	Pa s	$= \mathrm{kg\,m^{-1}\,s^{-1}}$
poise	P	$=10^{-1}$ Pa s
centipoise	cP	$=$ mPa s
kinematic viscosity, ν		
SI unit	$\mathrm{m^2\,s^{-1}}$	
stokes	St	$=10^{-4}\,\mathrm{m^2\,s^{-1}}$
thermodynamic temperature, T		
kelvin (SI unit)	K	
degree Rankine[2]	°R	$=(5/9)$ K
entropy, S		
heat capacity, C		
SI unit	$\mathrm{J\,K^{-1}}$	
clausius	Cl	$= cal_{th}/\mathrm{K} = 4.184\,\mathrm{J\,K^{-1}}$

(2) $T/°\mathrm{R} = (9/5)\ T/\mathrm{K}$. Also, Celsius temperature θ is related to thermodynamic temperature T by the equation

$$\theta/°\mathrm{C} = T/\mathrm{K} - 273.15.$$

Similarly Fahrenheit temperature θ_F is related to Celsius temperature θ by the equation

$$\theta_F/°\mathrm{F} = (9/5)\,(\theta/°\mathrm{C}) + 32.$$

Name	*Symbol*	*Relation to SI*
molar entropy, S_m		
molar heat capacity, C_m		
SI unit	$J\,K^{-1}\,mol^{-1}$	
entropy unit	e.u.	$=cal_{th}\,K^{-1}\,mol^{-1}=4.184\,J\,K^{-1}\,mol^{-1}$
molar volume, V_m		
SI unit	$m^3\,mol^{-1}$	
amagat[3]	amagat	$=V_m$ of real gas at 1 atm and 273.15 K, $\approx 22.414\times 10^{-3}\,m^3\,mol^{-1}$
molar density, $1/V_m$		
SI unit	$mol\,m^{-3}$	
amagat[3]	amagat	$=1/V_m$ for a real gas at 1 atm and 273.15 K, $\approx 44.615\,mol\,m^{-3}$
plane angle, α		
radian (SI unit)	rad	
degree	°	$=rad\times 2\pi/360\approx(1/57.295\,78)\,rad$
minute	′	$=degree/60$
second	″	$=degree/3600$
grade	grad	$=rad\times 2\pi/400\approx(1/63.661\,98)\,rad$
radioactivity, A		
becquerel (SI unit)	Bq	$=s^{-1}$
curie	Ci	$=3.7\times 10^{10}\,Bq$
absorbed dose of radiation[4]		
gray (SI unit)	Gy	$=J\,kg^{-1}$
rad	rad	$=0.01\,Gy$
dose equivalent		
sievert (SI unit)	Sv	$=J\,kg^{-1}$
rem	rem	$\approx 0.01\,Sv$
electric current, I		
ampere (SI unit)	A	
esu, Gau	$(10/\zeta)$ A	$\approx 3.335\,64\times 10^{-10}\,A$
emu (biot)	Bi	$=10\,A$
au	$eE_h/\hbar$	$\approx 6.623\,62\times 10^{-3}\,A$
electric charge, Q		
coulomb (SI unit)	C	$=A\,s$
franklin (esu, Gau)	Fr	$=(10/\zeta)\,C\approx 3.335\,64\times 10^{-10}\,C$
emu (abcoulomb)		$=10\,C$
proton charge (au)	e	$\approx 1.602\,18\times 10^{-19}\,C\approx 4.803\,21\times 10^{-10}\,Fr$

(3) The name 'amagat' is unfortunately used as a unit for both molar volume and molar density. Its value is slightly different for different gases, reflecting the deviation from ideal behaviour for the gas being considered.
(4) The unit röntgen, employed to express exposure of X or γ radiations, is equal to: $R=2.58\times 10^{-4}\,C\,kg^{-1}$.

Name	*Symbol*	*Relation to SI*
Charge density, ρ		
SI unit	$\mathrm{C\,m^{-3}}$	
esu, Gau	$\mathrm{Fr\,cm^{-3}}$	$=10^{7}\,\zeta^{-1}\,\mathrm{C\,m^{-3}}$
		$\approx 3.335\,64\times10^{-4}\,\mathrm{C\,m^{-3}}$
au	ea_0^{-3}	$\approx 1.08120\times10^{-12}\,\mathrm{C\,m^{-3}}$
electric potential, V, ϕ		
volt (SI unit)	V	$=\mathrm{J\,C^{-1}}=\mathrm{J\,A^{-1}\,s^{-1}}$
esu, Gau	$\mathrm{erg\,Fr^{-1}}$	$=\mathrm{Fr\,cm^{-1}}/4\pi\varepsilon_0=299.792\,458\ \mathrm{V}$
$\mathrm{cm^{-1}}$ (footnote 5)	$e\,\mathrm{cm^{-1}}/4\pi\varepsilon_0$	$\approx 1.439\,97\times10^{-7}\ \mathrm{V}$
au	$e/4\pi\varepsilon_0 a_0$	$=E_\mathrm{h}/e\approx 27.2114\ \mathrm{V}$
mean international volt		$=1.000\,34\ \mathrm{V}$
US international volt		$=1.000\,330\ \mathrm{V}$
electric resistance, R		
ohm (SI unit)	Ω	$=\mathrm{V\,A^{-1}}=\mathrm{m^2\,kg\,s^{-3}\,A^{-2}}$
mean international ohm		$=1.000\,49\ \Omega$
US international ohm		$=1.000\,495\ \Omega$
electric field, E		
SI unit	$\mathrm{V\,m^{-1}}$	$=\mathrm{J\,C^{-1}\,m^{-1}}$
esu, Gau	$\mathrm{Fr\,cm^{-2}}/4\pi\varepsilon_0$	$=2.997\,924\,58\times10^{4}\ \mathrm{V\,m^{-1}}$
$\mathrm{cm^{-2}}$ (footnote 5)	$e\,\mathrm{cm^{-2}}/4\pi\varepsilon_0$	$\approx 1.439\,97\times10^{-5}\ \mathrm{V\,m^{-1}}$
au	$e/4\pi\varepsilon_0 a_0^2$	$\approx 5.142\,21\times10^{11}\ \mathrm{V\,m^{-1}}$
electric field gradient, $E'_{\alpha\beta}, q_{\alpha\beta}$		
SI unit	$\mathrm{V\,m^{-2}}$	$=\mathrm{J\,C^{-1}\,m^{-2}}$
esu, Gau	$\mathrm{Fr\,cm^{-3}}/4\pi\varepsilon_0$	$=2.997\,924\,58\times10^{6}\ \mathrm{V\,m^{-2}}$
$\mathrm{cm^{-3}}$ (footnote 5)	$e\,\mathrm{cm^{-3}}/4\pi\varepsilon_0$	$\approx 1.439\,97\times10^{-3}\ \mathrm{V\,m^{-2}}$
au	$e/4\pi\varepsilon_0 a_0^3$	$\approx 9.717\,36\times10^{21}\ \mathrm{V\,m^{-2}}$
electric dipole moment, p, μ		
SI unit	$\mathrm{C\,m}$	
esu, Gau	$\mathrm{Fr\,cm}$	$\approx 3.335\,64\times10^{-12}\ \mathrm{C\,m}$
debye	D	$=10^{-18}\ \mathrm{Fr\,cm}\approx 3.335\,64\times10^{-30}\ \mathrm{C\,m}$
cm, dipole length[5]	$e\,\mathrm{cm}$	$\approx 1.602\,18\times10^{-21}\ \mathrm{C\,m}$
au	ea_0	$\approx 8.478\,36\times10^{-30}\ \mathrm{C\,m}$
electric quadrupole moment, $Q_{\alpha\beta}$, $\Theta_{\alpha\beta}$, eQ		
SI unit	$\mathrm{C\,m^2}$	
esu, Gau	$\mathrm{Fr\,cm^2}$	$\approx 3.335\,64\times10^{-14}\ \mathrm{C\,m^{-2}}$
$\mathrm{cm^2}$, quadrupole area[5]	$e\,\mathrm{cm^2}$	$\approx 1.602\,18\times10^{-23}\ \mathrm{C\,m^2}$
au	ea_0^2	$\approx 4.486\,55\times10^{-40}\ \mathrm{C\,m^2}$

Name	Symbol	Relation to SI
polarizability, α		
SI unit	$J^{-1}\,C^2\,m^2$	$=F\,m^2$
esu, Gau, cm^3, polarizability volume[5]	$4\pi\varepsilon_0\,cm^3$	$\approx 1.112\,65\times10^{-16}\,J^{-1}\,C^2\,m^2$
$Å^3$ (footnote 5)	$4\pi\varepsilon_0\,Å^3$	$\approx 1.112\,65\times10^{-40}\,J^{-1}\,C^2\,m^2$
au	$4\pi\varepsilon_0 a_0^3$	$\approx 1.648\,78\times10^{-41}\,J^{-1}\,C^2\,m^2$
electric displacement, D		
(*volume*) *polarization*, P		
SI unit	$C\,m^{-2}$	
esu, Gau	$Fr\,cm^{-2}$	$=(10^5/\zeta)\,C\,m^{-2}\approx 3.335\,64\times10^{-6}\,C\,m^{-2}$
(but note: the use of the esu or Gaussian unit for electric displacement usually implies that the irrational displacement is being quoted, $D^{(ir)}=4\pi D$. See section 7.4)		
magnetic flux density, B (*magnetic field*)		
tesla (SI unit)	T	$J\,A^{-1}\,m^{-2}=V\,s\,m^{-2}=Wb\,m^{-2}$
gauss (emu, Gau)	G	$=10^{-4}\,T$
au	$\hbar/ea_0^2$	$\approx 2.350\,52\times10^5\,T$
magnetic flux, Φ		
weber (SI unit)	Wb	$=J\,A^{-1}=V\,s$
maxwell (emu, Gau)	Mx	$=G\,cm^{-2}=10^{-8}\,Wb$
magnetic field, H		
(*volume*) *magnetization*, M		
SI unit	$A\,m^{-1}$	$=C\,s^{-1}\,m^{-1}$
oersted (emu, Gau)	Oe	$=10^3\,A\,m^{-1}$
(but note: in practice the oersted, Oe, is only used as a unit for $H^{(ir)}=4\pi H$; thus when $H^{(ir)}=1$ Oe, $H=(10^3/4\pi)\,A\,m^{-1}$; see section 7.4)		
magnetic dipole moment, m, μ		
SI unit	$A\,m^2$	$=J\,T^{-1}$
emu, Gau	$erg\,G^{-1}$	$=10\,A\,cm^2=10^{-3}\,J\,T^{-1}$
Bohr magneton[6]	μ_B	$\equiv e\hbar/2m_e\approx 9.274\,02\times10^{-24}\,J\,T^{-1}$
au	$e\hbar/m_e$	$=2\mu_B\approx 1.854\,80\times10^{-23}\,J\,T^{-1}$
nuclear magneton	μ_N	$=(m_e/m_p)\mu_B\approx 5.050\,79\times10^{-27}\,J\,T^{-1}$
magnetizability, ξ		
SI unit	$J\,T^{-2}$	$=C^2\,m^2\,kg^{-1}$
au	$e^2a_0^2/m_e$	$\approx 7.891\,04\times10^{-26}\,J\,T^{-2}$

(5) The units in quotation marks for electric potential through polarizability may be found in the literature, although they are strictly incorrect; they should be replaced in each case by the units given in the symbol column. Thus, for example, when a quadrupole moment is quoted in 'cm^2', the correct unit is $e\,cm^2$; and when a polarizability is quoted in '$Å^3$', the correct unit is $4\pi\varepsilon_0 Å^3$.

(6) The Bohr magneton μ_B is sometimes denoted BM (or B.M.), but this is not recommended.

Name	*Symbol*	*Relation to SI*
magnetic susceptibility, χ, κ		
SI unit	1	
emu, Gau	1	

(but note: in practice susceptibilities quoted in the context of emu or Gaussian units are always values for $\chi^{(ir)} = \chi/4\pi$; thus when $\chi^{(ir)} = 10^{-6}$, then $\chi = 4\pi \times 10^{-6}$; see section 7.3)

Name	*Symbol*	*Relation to SI*
molar susceptibility, χ_m		
SI unit	$m^3\ mol^{-1}$	
emu, Gau	$cm^3\ mol^{-1}$	$= 10^{-6}\ m^3\ mol^{-1}$

(but note: in practice the units $cm^3\ mol^{-1}$ usually imply that the irrational molar susceptibility is being quoted, $\chi_m^{(ir)} = \chi_m/4\pi$; thus for example if $\chi_m^{(ir)} = -15 \times 10^{-6}\ cm^3\ mol^{-1}$, which is often written as '-15 cgs ppm', then $\chi_m = -1.88 \times 10^{-10}\ m^3\ mol^{-1}$; see section 7.3)

7.3 THE esu, emu, GAUSSIAN, AND ATOMIC UNIT SYSTEMS

The SI equations of electromagnetic theory are usually used with physical quantities in SI units, in particular the four units m, kg, s, and A for length, mass, time and electric current. The basic equations for the electrostatic force between charges Q_1 and Q_2, and for the electromagnetic force between current elements $I_1 \mathrm{d}\boldsymbol{l}_1$ and $I_2 \mathrm{d}\boldsymbol{l}_2$, in vacuum, are written

$$\boldsymbol{F} = Q_1 Q_2 \boldsymbol{r}/4\pi\varepsilon_0 r^3, \tag{1a}$$

$$\boldsymbol{F} = (\mu_0/4\pi) I_1 \mathrm{d}\boldsymbol{l}_1 \times (I_2 \mathrm{d}\boldsymbol{l}_2 \times \boldsymbol{r})/r^3. \tag{1b}$$

The physical quantities ε_0 and μ_0, the permittivity and permeability of vacuum, respectively, have the values

$$\varepsilon_0 = (10^7/4\pi c_0{}^2)\ \mathrm{kg^{-1}m^{-1}C^2} \approx 8.854\,188 \times 10^{-12}\ \mathrm{C^2m^{-1}J^{-1}}, \tag{2a}$$

$$\mu_0 = 4\pi \times 10^{-7}\ \mathrm{N\,A^{-2}} \approx 1.256\,637 \times 10^{-6}\ \mathrm{N\,A^{-2}}. \tag{2b}$$

The value of μ_0 results from the definition of the ampere (section 3.2), which is such as to give μ_0 the value in (2b). The value of ε_0 then results from the relation

$$\varepsilon_0 \mu_0 = 1/c_0{}^2, \tag{3}$$

where c_0 is the speed of light in vacuum.

The numerical constants 4π are introduced into the definition of ε_0 and μ_0 because of the spherical symmetry involved in equations (1); in this way we avoid their appearance in later equations relating to systems with rectangular symmetry. When factors of 4π are introduced in this way, as in the SI, the equations are described as 'rationalized'. The alternative 'unrationalized' or 'irrational' form of the electromagnetic equations is discussed below.

Other systems of units and equations in common use in electromagnetic theory, in addition to the SI, are the esu system, the emu system, the Gaussian system, and the system of atomic units. The conversion from SI to these other systems may be understood in the following steps.

First, all of the alternative systems involve equations written in the irrational form, in place of the rationalized form used in the SI. This involves changes of factors of 4π, and the redefinition of certain physical quantities. Second, a particular choice of units is made in each case to give either ε_0 or μ_0 a simple chosen value. Third, in the case of the esu, emu, and Gaussian systems (but not in the case of atomic units) the system of four base units (and four independent dimensions) is dropped in favour of only three base units (and independent dimensions) by an appropriate choice of the definition of charge or current in terms of length, mass and time. All these changes are considered in more detail below. Finally, because of the complications resulting from the alternative choice of rational or irrational relations, and the alternative ways of choosing the base dimensions, the equations of electromagnetic theory are different in the different systems. These changes are summarized in table 7.4 which gives the conversion of equations between the SI and the alternative systems.

(i) *The change to irrational quantities and equations*

Equations (1) can be written in the alternative four-quantity irrational form by defining new quantities $\varepsilon_0^{(\mathrm{ir})}$ and $\mu_0^{(\mathrm{ir})}$, so that (1a,b) become:

$$\boldsymbol{F} = Q_1 Q_2 \boldsymbol{r}/\varepsilon_0^{(\mathrm{ir})} r^3, \tag{4a}$$

$$\boldsymbol{F} = \mu_0^{(\mathrm{ir})} I_1 \mathrm{d}\boldsymbol{l}_1 \times (I_2 \mathrm{d}\boldsymbol{l}_2 \times \boldsymbol{r})/r^3. \tag{4b}$$

The new quantities are related to ε_0 and μ_0 by the equations:

$$\varepsilon_0^{(\mathrm{ir})} = 4\pi\varepsilon_0, \tag{5a}$$

$$\mu_0^{(\mathrm{ir})} = \mu_0/4\pi. \tag{5b}$$

When the equations of electromagnetic theory are written in this alternative irrational form, six other new quantities are defined in addition to $\varepsilon_0^{(\mathrm{ir})}$ and $\mu_0^{(\mathrm{ir})}$; namely $\varepsilon^{(\mathrm{ir})}$, $\mu^{(\mathrm{ir})}$, $D^{(\mathrm{ir})}$, $H^{(\mathrm{ir})}$, $\chi_\mathrm{e}^{(\mathrm{ir})}$ (the electric susceptibility), and $\chi^{(\mathrm{ir})}$ (the magnetic susceptibility). The definitions of other quantities remain unchanged. In each case we denote the new quantities by a superscript (ir), for irrational. The new quantities are defined in terms of the old quantities by the equations:

$$\varepsilon^{(\mathrm{ir})} = 4\pi\varepsilon, \qquad (6a)$$

$$\mu^{(\mathrm{ir})} = \mu/4\pi, \qquad (6b)$$

$$D^{(\mathrm{ir})} = 4\pi D, \qquad (7a)$$

$$H^{(\mathrm{ir})} = 4\pi H, \qquad (7a)$$

$$\chi_\mathrm{e}^{(\mathrm{ir})} = \chi_\mathrm{e}/4\pi, \qquad (8a)$$

$$\chi^{(\mathrm{ir})} = \chi/4\pi. \qquad (8b)$$

All of the equations of electromagnetic theory can now be transformed from the SI into the irrational form by using equations (5a,b), (6a,b), (7a,b) and (8a,b) to eliminate ε_0, μ_0, ε, μ, D, H, χ_e, and χ from the SI equations in favour of the corresponding irrational quantities distinguished by a superscript (ir).

The notation of a superscript (ir), used here to distinguish irrational quantities from their rational counterparts, where the definitions differ, is clumsy. However, in the published literature it is unfortunately customary to use exactly the same symbol for the quantities ε, μ, D, H, χ_e, and χ whichever definition (and corresponding set of equations) is in use. It is as though atomic and molecular physicists were to use the same symbol h for Planck's constant, and Planck's constant/2π. Fortunately the different symbols h and $\hbar$ have been adopted in this case, and so we are able to write equations like $h = 2\pi\hbar$. Without some distinction in the notation, equations like (5), (6), (7) and (8) are impossible to write, and it is then difficult to discuss the relations between the rationalized SI equations and quantities, and their irrational esu and emu equivalents. This is the reason for the rather cumbersome notation adopted here to distinguish quantities defined by different equations in the different systems.

(ii) *The esu system*

The esu system is based on irrational equations and quantities, and may be described either in terms of four base units and four independent dimensions, or—as is more usual—in terms of three base units and three independent dimensions.

When four base units are used, they are taken to be the cm, g, and s for length, mass and time, and the franklin[1] (symbol Fr) for the esu of charge, 1 Fr being chosen to be of such a magnitude that $\varepsilon_0^{(\mathrm{ir})} = 1\ \mathrm{Fr}^2/\mathrm{erg\ cm}$. An equivalent definition of the franklin is that two charges of 1 Fr, 1 cm apart in a vacuum, repel each other with a force of one dyne. Other units are then derived from these four by the usual rules for constructing a coherent set of units from a set of base units.

The alternative and more usual form of the esu system is built on only three base units and three independent dimensions. This is achieved by defining the dimension of charge to be the same as that of $[(\text{energy}) \times (\text{length})]^{1/2}$, so that $1\ \mathrm{Fr}^2 = 1\ \mathrm{erg\ cm}$. The Fr then disappears as a unit, and the constant $\varepsilon_0^{(\mathrm{ir})}$ is dimensionless, and equal to 1, so that it may be omitted from all equations. Thus equation (4a) for the force between charges in vacuum, for example, becomes simply

$$\boldsymbol{F} = Q_1 Q_2 \boldsymbol{r}/r^3. \qquad (9)$$

---

(1) The name 'franklin', symbol Fr, for the esu of charge was suggested by Guggenheim more than 40 years ago (*Nature*, **148** (1941) 751). Although it has not been widely adopted, this name and symbol are used here for convenience as a compact expression for the esu of charge. The name 'statcoulomb' has also been used for the esu of charge.

This also means that the permittivity of a dielectric medium, $\varepsilon^{(\mathrm{ir})}$, is exactly the same as the relative permittivity or dielectric constant ε_r, so that only one of these quantities is required—which is usually simply called the permittivity, ε. Finally since $\varepsilon_0^{(\mathrm{ir})}=1$, equations (3) and (5) require that $\mu_0^{(\mathrm{ir})}=1/c_0{}^2$.

To summarize, the transformation of equations from the four-quantity SI to the three-quantity esu system is achieved by making the substitutions $\varepsilon_0=1/4\pi$, $\mu_0=4\pi/c_0{}^2$, $\varepsilon=\varepsilon_r/4\pi$, $D=D^{(\mathrm{ir})}/4\pi$, and $\chi_e=4\pi\chi_e^{(\mathrm{ir})}$.

(iii) *The emu system*

The emu system is also based on irrational equations and quantities, and may similarly be described in terms of either four or three base units.

When described in terms of four base units, they are taken as the cm, g, s, and the unit of electric current, which we call the (emu of current). This is chosen to be of such a magnitude that $\mu_0^{(\mathrm{ir})}=1$ cm g s^{-2} (emu of current)$^{-2}$. (An equivalent definition of the emu of current is that the force between two parallel wires, 1 cm apart in a vacuum, each carrying 1 emu of current, is 2 dyn per cm of wire.) Comparison with the definition of the ampere then shows that 1 (emu of current) = 10 A. Other units are derived from these four by the usual rules.[2]

In the more usual description of the emu system only three base units and three independent dimensions are used. The dimension (electric current) is defined to be the same as that of (force)$^{1/2}$, so that 1 (emu of current)^2 = 1 g cm s^{-2} = 1 dyn. The (emu of current) then disappears as a unit, and the constant $\mu_0^{(\mathrm{ir})}$ is dimensionless and equal to 1, so that it may be omitted from all equations. Thus equation (4b) for the force between current elements in vacuum, for example, becomes simply

$$\boldsymbol{F}=I_1\mathrm{d}\boldsymbol{l}_1\times(I_2\mathrm{d}\boldsymbol{l}_2\times\boldsymbol{r})/r^3. \qquad (10)$$

The permeability of a magnetic medium $\mu^{(\mathrm{ir})}$ is identical to the relative permeability or magnetic constant μ_r, and is simply called the permeability. Finally $\varepsilon_0^{(\mathrm{ir})}=1/c_0{}^2$ in the emu system.

To summarize, the transformation from the four-quantity SI to the three-quantity emu system is achieved by making the substitutions $\mu_0=4\pi$, $\varepsilon_0=1/4\pi c_0{}^2$, $\mu=4\pi\mu_r$, $H=H^{(\mathrm{ir})}/4\pi$, and $\chi=4\pi\chi^{(\mathrm{ir})}$.

(iv) *The Gaussian system*

The Gaussian system is a mixture of the esu system and the emu system, expressed in terms of three base units, esu being used for quantities in electrostatics and emu for electrodynamics. It is thus a hybrid system, and this gives rise to complications in both the equations and the units.

In the usual form of the Gaussian system, the following quantities are defined as in the esu system: charge Q, current I, electric field $\boldsymbol{E}$, electric displacement $\boldsymbol{D}^{(\mathrm{ir})}$, electric potential V, polarization $\boldsymbol{P}$, electric dipole $\boldsymbol{p}$, electric susceptibility $\chi_e^{(\mathrm{ir})}$, polarizability α, and capacitance C.

The following quantities are defined as in the emu system: magnetic flux density $\boldsymbol{B}$, magnetic flux Φ, magnetic potential $\boldsymbol{A}$, magnetic field $\boldsymbol{H}^{(\mathrm{ir})}$, magnetization $\boldsymbol{M}$, magnetic susceptibility $\chi^{(\mathrm{ir})}$, magnetic dipole $\boldsymbol{m}$, and magnetizability ξ. Neither $\varepsilon_0^{(\mathrm{ir})}$ nor $\mu_0^{(\mathrm{ir})}$ appear in the Gaussian equations, both being set equal to 1; the permittivity $\varepsilon^{(\mathrm{ir})}=\varepsilon_r$, and the permeability $\mu^{(\mathrm{ir})}=\mu_r$. However, the effect of equation (3) is that each physical quantity in the esu system differs in magnitude and dimensions from the corresponding emu quantity by some power of c_0. Thus the conversion of each SI equation of electromagnetic theory into the Gaussian form introduces factors of c_0, which are required to ensure internal consistency.

The transformation of the more important equations between the Gaussian system and the SI is given in table 7.4 below.

(2) The name biot, symbol Bi, has been used for the (emu of current).

(v) *Atomic units* [8] (see also section 3.8, p.70)

The so-called 'atomic units' are fundamental constants (and combinations of such constants) that arise in atomic and molecular electronic structure calculations, which are conveniently treated as though they were units. They may be regarded as a coherent system of units built on combinations of the four independent dimensions length, mass, time, and electric charge. (The remaining dimensions used in the SI do not appear in electronic structure calculations.) The electromagnetic equations are taken in the irrational form. Atomic units are defined by taking the base unit of mass to be the electron rest mass m_e, the base unit of charge to be the elementary charge e, the base unit of action (angular momentum) to be $\hbar = h/2\pi$ (where h is the Planck constant), and by choosing the base units of length a_0 and of energy E_h to be coherent with m_e, e and $\hbar$ in such a way that $4\pi\varepsilon_0 = \varepsilon_0^{(ir)} = e^2/E_h a_0$. (In a similar way the unit of charge in the four-quantity esu system, the franklin, Fr, is chosen to give $4\pi\varepsilon_0$ a simple value: $4\pi\varepsilon_0 = 1\ \mathrm{Fr}^2/\mathrm{erg\ cm}$.)

The atomic unit of energy E_h is called the hartree; it is (approximately) twice the ionization energy of a hydrogen atom in its 1s ground state. The atomic unit of length a_0 is called the bohr; it is (approximately) the distance of maximum radial density from the nucleus in the 1s orbital of a hydrogen atom. Clearly only four of the five units m_e, e, $\hbar$, E_h and a_0 can be independent; useful ways of writing the interrelation are:

$$E_h = \hbar^2/m_e a_0{}^2 = e^2/4\pi\varepsilon_0 a_0 = m_e e^4/(4\pi\varepsilon_0)^2\hbar^2. \tag{11}$$

Conversion factors from atomic units to the SI are included in table 7.2 (p.104), and the five atomic units which have special names and symbols (described above), as well as a number of other atomic units, are also listed in table 3.8 (p.70).

The importance of atomic units lies in the fact that *ab initio* calculations in theoretical chemistry necessarily give results in atomic units (i.e. as multiples of m_e, e, $\hbar$, E_h and a_0). They are sometimes described as the 'natural units' of electronic calculations in theoretical chemistry. Indeed the results of such calculations can only be converted to other units (such as the SI) by using the current best estimates of the physical constants m_e, e, $\hbar$, etc., themselves expressed in SI units. It is thus appropriate for theoretical chemists to express their results in atomic units, and for the reader to convert to other units as and when necessary. This is also the reason why atomic units are written in italic (sloping) type rather than in the roman (upright) type usually used for units: the atomic units are physical quantities chosen from the fundamental physical constants of electronic structure calculations. There is, however, no authority from CGPM for designating these quantities as 'units', despite the fact that they are treated as units and called 'atomic units' by workers in the field.

Some authors who use atomic units use the customary symbols for physical quantities to represent the numerical values of quantities in the form (*physical quantity*)/(*atomic unit*), so that all quantities appear as pure numbers. Thus for example the Schrödinger equation for the hydrogen atom is written in SI in the form:

$$-(\hbar^2/2m_e)\nabla_r^2\psi - (e^2/4\pi\varepsilon_0 r)\psi = E\psi, \tag{12}$$

where ∇_r denotes derivatives with respect to r. After dividing throughout by E_h and making use of (11), this becomes

$$-\tfrac{1}{2}a_0{}^2\nabla_r^2\psi - (a_0/r)\psi = (E/E_h)\psi. \tag{13}$$

If we now define $\rho = r/a_0$, and $E' = E/E_h$, so that ρ and E' are dimensionless numbers giving the numerical values of r and E in atomic units, then (13) can be written

$$-\tfrac{1}{2}\nabla_\rho^2\psi - (1/\rho)\psi = E'\psi, \tag{14}$$

where ∇_ρ denotes derivatives with respect to ρ. Equation (14), in which each coefficient of ψ is dimensionless, is commonly described as being 'expressed in atomic units', and is the form usually adopted by theoretical chemists. Although the power of dimensional analysis is lost in this form, the symbolism has the advantage of simplicity. In using this form it is helpful to distinguish the dimensionless quantities which are here denoted ρ and E' from the customary physical quantities r and E themselves, but many authors make no distinction in either the symbol or the name.

Some authors also use the symbol 'au' (or 'a.u.') for every atomic unit, in place of the appropriate combination of the explicit symbols m_e, e, $\hbar$, E_h and a_0. This should be avoided. Appropriate combinations of m_e, e, $\hbar$, E_h and a_0 for the atomic unit of various physical quantities are given in tables 3.8 (p.70) and 7.2 (p.104).

Examples $E = -0.345\ E_h$, not -0.345 atomic units
$r = 1.567\ a_0$, not 1.567 a.u. or 1.567 au

7.4 TRANSFORMATION OF EQUATIONS OF ELECTROMAGNETIC THEORY BETWEEN THE SI, THE FOUR-QUANTITY IRRATIONAL FORM, AND THE GAUSSIAN FORM

Note that the esu equations may be obtained from the four-quantity irrational equations by putting $\varepsilon_0^{(ir)}=1$, and $\mu_0^{(ir)}=1/c_0{}^2$; the emu equations may be obtained by putting $\mu_0^{(ir)}=1$, and $\varepsilon_0^{(ir)}=1/c_0{}^2$.

SI relation	*Four-quantity irrational relation*	*Gaussian relation*
force on a moving charge Q with velocity $\boldsymbol{v}$:		
$\boldsymbol{F}=Q(\boldsymbol{E}+\boldsymbol{v}\times\boldsymbol{B})$	$\boldsymbol{F}=Q(\boldsymbol{E}+\boldsymbol{v}\times\boldsymbol{B})$	$\boldsymbol{F}=Q(\boldsymbol{E}+\boldsymbol{v}\times\boldsymbol{B}/c_0)$
force between charges in vacuum:		
$\boldsymbol{F}=Q_1Q_2\boldsymbol{r}/4\pi\varepsilon_0 r^3$	$\boldsymbol{F}=Q_1Q_2\boldsymbol{r}/\varepsilon_0^{(ir)}r^3$	$\boldsymbol{F}=Q_1Q_2\boldsymbol{r}/r^3$
potential around a charge in vacuum:		
$V=Q/4\pi\varepsilon_0 r$	$V=Q/\varepsilon_0^{(ir)}r$	$V=Q/r$
relation between field and potential:		
$\boldsymbol{E}=-\text{grad}\,V$	$\boldsymbol{E}=-\text{grad}\,V$	$\boldsymbol{E}=-\text{grad}\,V$
field due to a charge distribution in vacuum:		
$\text{div}\,\boldsymbol{E}=\rho/\varepsilon_0$	$\text{div}\,\boldsymbol{E}=4\pi\rho/\varepsilon_0^{(ir)}$	$\text{div}\,\boldsymbol{E}=4\pi\rho$
capacitance of a parallel plate condenser, area A, separation d:		
$C=\varepsilon_0\varepsilon_r A/d$	$C=\varepsilon_0^{(ir)}\varepsilon_r A/4\pi d$	$C=\varepsilon_r A/4\pi d$
electric dipole moment of a charge distribution:		
$\boldsymbol{p}=\int\rho\boldsymbol{r}\,\mathrm{d}V$	$\boldsymbol{p}=\int\rho\boldsymbol{r}\,\mathrm{d}V$	$\boldsymbol{p}=\int\rho\boldsymbol{r}\,\mathrm{d}V$
potential around a dipole in vacuum:		
$V=\boldsymbol{p}\cdot\boldsymbol{r}/4\pi\varepsilon_0 r^3$	$V=\boldsymbol{p}\cdot\boldsymbol{r}/\varepsilon_0^{(ir)}r^3$	$V=\boldsymbol{p}\cdot\boldsymbol{r}/r^3$
energy of a charge distribution in an electric field:		
$E_p=QV-\boldsymbol{p}\cdot\boldsymbol{E}+\cdots$	$E_p=QV-\boldsymbol{p}\cdot\boldsymbol{E}+\cdots$	$E_p=QV-\boldsymbol{p}\cdot\boldsymbol{E}+\cdots$
electric dipole moment induced by a field:		
$\boldsymbol{p}=\alpha\boldsymbol{E}+\cdots$	$\boldsymbol{p}=\alpha\boldsymbol{E}+\cdots$	$\boldsymbol{p}=\alpha\boldsymbol{E}+\cdots$
relations between $\boldsymbol{E}$, $\boldsymbol{D}$ and $\boldsymbol{P}$:		
$\boldsymbol{E}=(\boldsymbol{D}-\boldsymbol{P})/\varepsilon_0$	$\boldsymbol{E}=(\boldsymbol{D}^{(ir)}-4\pi\boldsymbol{P})/\varepsilon_0^{(ir)}$	$\boldsymbol{E}=\boldsymbol{D}^{(ir)}-4\pi\boldsymbol{P}$
$\boldsymbol{E}=\boldsymbol{D}/\varepsilon_0\varepsilon_r$	$\boldsymbol{E}=\boldsymbol{D}^{(ir)}/\varepsilon_0^{(ir)}\varepsilon_r$	$\boldsymbol{E}=\boldsymbol{D}^{(ir)}/\varepsilon_r$
relations involving the electric susceptibility:		
$\varepsilon_r=1+\chi_e$	$\varepsilon_r=1+4\pi\chi_e^{(ir)}$	$\varepsilon_r=1+4\pi\chi_e^{(ir)}$
$\boldsymbol{P}=\chi_e\varepsilon_0\boldsymbol{E}$	$\boldsymbol{P}=\chi_e^{(ir)}\varepsilon_0^{(ir)}\boldsymbol{E}$	$\boldsymbol{P}=\chi_e^{(ir)}\boldsymbol{E}$
force between current elements in vacuum:		
$\boldsymbol{F}=\dfrac{\mu_0}{4\pi}\dfrac{I\mathrm{d}\boldsymbol{l}_1\times(I\mathrm{d}\boldsymbol{l}_2\times\boldsymbol{r})}{r^3}$	$\boldsymbol{F}=\dfrac{\mu_0^{(ir)}I\mathrm{d}\boldsymbol{l}_1\times(I\mathrm{d}\boldsymbol{l}_2\times\boldsymbol{r})}{r^3}$	$\boldsymbol{F}=\dfrac{I\mathrm{d}\boldsymbol{l}_1\times(I\mathrm{d}\boldsymbol{l}_2\times\boldsymbol{r})}{c_0{}^2r^3}$
force on a current element in a field:		
$\boldsymbol{F}=I\mathrm{d}\boldsymbol{l}\times\boldsymbol{B}$	$\boldsymbol{F}=I\mathrm{d}\boldsymbol{l}\times\boldsymbol{B}$	$\boldsymbol{F}=I\mathrm{d}\boldsymbol{l}\times\boldsymbol{B}/c_0$

SI relation	Four-quantity irrational relation	Gaussian relation
potential due to a current element in vacuum:		
$A=(\mu_0/4\pi)(I\mathrm{d}\boldsymbol{l}/r)$	$A=\mu_0^{(\mathrm{ir})}I\mathrm{d}\boldsymbol{l}/r$	$A=I\mathrm{d}\boldsymbol{l}/c_0 r$
relation between field and potential:		
$\boldsymbol{B}=\mathrm{curl}\,A$	$\boldsymbol{B}=\mathrm{curl}\,A$	$\boldsymbol{B}=\mathrm{curl}\,A$
field due to a current element in vacuum:		
$\boldsymbol{B}=(\mu_0/4\pi)(I\mathrm{d}\boldsymbol{l}\times\boldsymbol{r}/r^3)$	$\boldsymbol{B}=\mu_0^{(\mathrm{ir})}I\mathrm{d}\boldsymbol{l}\times\boldsymbol{r}/r^3$	$\boldsymbol{B}=I\mathrm{d}\boldsymbol{l}\times\boldsymbol{r}/c_0 r^3$
field due to a current density $\boldsymbol{j}$ in vacuum:		
$\mathrm{curl}\,\boldsymbol{B}=\mu_0\boldsymbol{j}$	$\mathrm{curl}\,\boldsymbol{B}=4\pi\mu_0^{(\mathrm{ir})}\boldsymbol{j}$	$\mathrm{curl}\,\boldsymbol{B}=4\pi\boldsymbol{j}/c_0$
magnetic dipole of a current loop of area $\mathrm{d}\boldsymbol{A}$:		
$\boldsymbol{m}=I\mathrm{d}A$	$\boldsymbol{m}=I\mathrm{d}A$	$\boldsymbol{m}=I\mathrm{d}A/c_0$
potential around a magnetic dipole in vacuum:		
$A=(\mu_0/4\pi)(\boldsymbol{m}\times\boldsymbol{r}/r^3)$	$A=\mu_0^{(\mathrm{ir})}\boldsymbol{m}\times\boldsymbol{r}/r^3$	$A=\boldsymbol{m}\times\boldsymbol{r}/c_0 r^3$
energy of a magnetic dipole in a field:		
$E_\mathrm{p}=-\boldsymbol{m}\cdot\boldsymbol{B}$	$E_\mathrm{p}=-\boldsymbol{m}\cdot\boldsymbol{B}$	$E_\mathrm{p}=-\boldsymbol{m}\cdot\boldsymbol{B}$
magnetic dipole induced by a field:		
$\boldsymbol{m}=\xi\boldsymbol{B}$	$\boldsymbol{m}=\xi\boldsymbol{B}$	$\boldsymbol{m}=\xi\boldsymbol{B}$
relations between **B**, **H** and **M**:		
$\boldsymbol{B}=\mu_0(\boldsymbol{H}+\boldsymbol{M})$	$\boldsymbol{B}=\mu_0^{(\mathrm{ir})}(\boldsymbol{H}^{(\mathrm{ir})}+4\pi\boldsymbol{M})$	$\boldsymbol{B}=\boldsymbol{H}^{(\mathrm{ir})}+4\pi\boldsymbol{M}$
$\boldsymbol{B}=\mu_0\mu_\mathrm{r}\boldsymbol{H}$	$\boldsymbol{B}=\mu_0^{(\mathrm{ir})}\mu_\mathrm{r}\boldsymbol{H}^{(\mathrm{ir})}$	$\boldsymbol{B}=\mu_\mathrm{r}\boldsymbol{H}^{(\mathrm{ir})}$
relations involving the magnetic susceptibility:		
$\mu_\mathrm{r}=1+\chi$	$\mu_\mathrm{r}=1+4\pi\chi^{(\mathrm{ir})}$	$\mu_\mathrm{r}=1+4\pi\chi^{(\mathrm{ir})}$
$\boldsymbol{M}=\chi\boldsymbol{B}/\mu_0$	$\boldsymbol{M}=\chi^{(\mathrm{ir})}\boldsymbol{B}/\mu_0^{(\mathrm{ir})}$	$\boldsymbol{M}=\chi^{(\mathrm{ir})}\boldsymbol{B}$
Curie relation:		
$\chi_\mathrm{m}=V_\mathrm{m}\chi$	$\chi_\mathrm{m}^{(\mathrm{ir})}=V_\mathrm{m}\chi^{(\mathrm{ir})}$	$\chi_\mathrm{m}^{(\mathrm{ir})}=V_\mathrm{m}\chi^{(\mathrm{ir})}$
$=L\mu_0 m^2/3kT$	$=L\mu_0^{(\mathrm{ir})}m^2/3kT$	$=Lm^2/3kT$
Maxwell equations:		
$\mathrm{div}\,\boldsymbol{D}=\rho$	$\mathrm{div}\,\boldsymbol{D}^{(\mathrm{ir})}=4\pi\rho$	$\mathrm{div}\,\boldsymbol{D}^{(\mathrm{ir})}=4\pi\rho$
$\mathrm{div}\,\boldsymbol{B}=0$	$\mathrm{div}\,\boldsymbol{B}=0$	$\mathrm{div}\,\boldsymbol{B}=0$
$\mathrm{curl}\,\boldsymbol{E}+\partial\boldsymbol{B}/\partial t=0$	$\mathrm{curl}\,\boldsymbol{E}+\partial\boldsymbol{B}/\partial t=0$	$\mathrm{curl}\,\boldsymbol{E}+\partial\boldsymbol{B}/\partial t=0$
$\mathrm{curl}\,\boldsymbol{H}-\partial\boldsymbol{D}/\partial t=0$	$\mathrm{curl}\,\boldsymbol{H}^{(\mathrm{ir})}-\dfrac{\partial\boldsymbol{D}^{(\mathrm{ir})}}{\partial t}=0$	$\mathrm{curl}\,\boldsymbol{H}^{(\mathrm{ir})}-\dfrac{\partial\boldsymbol{D}^{(\mathrm{ir})}}{\partial t}=0$
energy density of radiation:		
$U/V=(\boldsymbol{E}\cdot\boldsymbol{D}+\boldsymbol{B}\cdot\boldsymbol{H})/2$	$U/V=\dfrac{\boldsymbol{E}\cdot\boldsymbol{D}^{(\mathrm{ir})}+\boldsymbol{B}\cdot\boldsymbol{H}^{(\mathrm{ir})}}{8\pi}$	$U/V=\dfrac{\boldsymbol{E}\cdot\boldsymbol{D}^{(\mathrm{ir})}+\boldsymbol{B}\cdot\boldsymbol{H}^{(\mathrm{ir})}}{8\pi}$
rate of energy flow (Poynting vector):		
$\boldsymbol{S}=\boldsymbol{E}\times\boldsymbol{H}$	$\boldsymbol{S}=\boldsymbol{E}\times\boldsymbol{H}^{(\mathrm{ir})}/4\pi$	$\boldsymbol{S}=c_0\boldsymbol{E}\times\boldsymbol{H}^{(\mathrm{ir})}/4\pi$

8
References

8.1 PRIMARY SOURCES

1 IUPAC-Physical Chemistry Division
Manual of Symbols and Terminology for Physicochemical Quantities and Units
(a) 1st ed. *Pure Appl. Chem.* **21** (1970) 1–38.
(b) 2nd ed. Butterworths, London 1975.
(c) 3rd ed. *Pure Appl. Chem.* **51** (1979) 1–36.
(d) Appendix I—Definitions of Activities and Related Quantities, *Pure Appl. Chem.* **51** (1979) 37–41.
(e) Appendix II—Definitions, Terminology and Symbols in Colloid and Surface Chemistry, Part I, *Pure Appl. Chem.* **31** (1972) 577–638.
(f) Section 1.13: Selected Definitions, Terminology and Symbols for Rheological Properties, *Pure Appl. Chem.* **51** (1979) 1213–1218.
(g) Section 1.14: Light Scattering, *Pure Appl. Chem.* **55** (1983) 931–941.
(h) Part II: Heterogeneous Catalysis, *Pure Appl. Chem.* **46** (1976) 71–90.
(i) Appendix III—Electrochemical Nomenclature, *Pure Appl. Chem.* **37** (1974) 499–516.
(j) Appendix IV—Notation for States and Processes, Significance of the Word 'Standard' in Chemical Thermodynamics, and Remarks on Commonly Tabulated Forms of Thermodynamic Functions, *Pure Appl. Chem.* **54** (1982) 1239–1250.
(k) Appendix V—Symbolism and Terminology in Chemical Kinetics, *Pure Appl. Chem.* **53** (1981) 753–771.

2 Bureau International des Poids et Mesures, Le Système International d'Unités (SI), 5th French and English Edition, BIPM, Sèvres 1985.

3 IUPAP-SUN, Symbols, Units and Nomenclature in Physics, Document U.I.P. 20, *Physica* **93A** (1978) 1–60.

4 International Standards ISO, International Organization for Standardization, Geneva.
(a) ISO 31/0–1981, General Principles Concerning Quantities, Units and Symbols.
(b) ISO 31/1–1978, Quantities and Units of Space and Time.
(c) ISO 31/2–1978, Quantities and Units of Periodic and Related Phenomena.
(d) ISO 31/3–1978, Quantities and Units of Mechanics.
(e) ISO 31/4–1978, Quantities and Units of Heat.
(f) ISO 31/5–1979, Quantities and Units of Electricity and Magnetism.
(g) ISO 31/6–1980, Quantities and Units of Light and Related Electromagnetic Radiations.
(h) ISO 31/7–1978, Quantities and Units of Acoustics.
(i) ISO 31/8–1980, Quantities and Units of Physical Chemistry and Molecular Physics.
(j) ISO 31/9–1980, Quantities and Units of Atomic and Nuclear Physics.
(k) ISO 31/10–1980, Quantities and Units of Nuclear Reactions and Ionizing Radiations.
(l) ISO 31/11–1978, Mathematical Signs and Symbols for Use in Physical Sciences and Technology.
(m) ISO 31/12–1982, Dimensionless Parameters.
(n) ISO 31/13–1981, Quantities and Units of Solid State Physics.

5 ISO 1000–1981, SI Units and Recommendations for the Use of Their Multiples and of Certain Other Units.
All the standards listed here (4-5) are jointly reproduced in the ISO Standards Handbook 2, *Units of Measurement*, ISO, Geneva 1982.

6 ISO 2955–1983, Information Processing—Representations of SI and Other Units for Use in Systems with Limited Character Sets.

8.2 IUPAC REFERENCES

7 Use of Abbreviations in the Chemical Literature, *Pure Appl. Chem.* **52** (1980) 2229–2232.

8 Expression of Results in Quantum Chemistry, *Pure Appl. Chem.* **50** (1978) 75–79.

9 Recommendations for Presentation of Infrared Absorption Spectra in Data Collections: A—Condensed Phases, *Pure Appl. Chem.* **50** (1978) 231–236.

10 Presentation of Raman Spectra in Data Collections, *Pure Appl. Chem.* **53** (1981) 1879–1885.

11 Recommendations for the Presentation on NMR Data for Publication in Chemical Journals, *Pure Appl. Chem.* **29** (1972) 625–628.

12 Presentation of NMR Data for Publication in Chemical Journals: B—Conventions Relating to Spectra from Nuclei other than Protons, *Pure Appl. Chem.* **45** (1976) 217–219.

13 Nomenclature and Spectral Presentation in Electron Spectroscopy Resulting from Excitation by Photons, *Pure Appl. Chem.* **45** (1976) 221–224.

14 Nomenclature and Conventions for Reporting Mössbauer Spectroscopic Data, *Pure Appl. Chem.* **45** (1976) 211–216.

15 Recommendations for Symbolism and Nomenclature for Mass Spectroscopy. *Pure Appl. Chem.* **50** (1978) 65–73.

16 Definition and Symbolism of Molecular Force Constants, *Pure Appl. Chem.* **50** (1978) 1707–1713.

17 Names, Symbols, Definitions and Units of Quantities in Optical Spectroscopy. *Pure Appl. Chem.* **57** (1985) 105–120.

18 Nomenclature, Symbols, Units and their Usage in Spectrochemical Analysis. I: General Atomic Emission Spectroscopy, *Pure Appl. Chem.* **30** (1972) 651–679.

19 Nomenclature, Symbols, Units and their Usage in Spectrochemical Analysis. VI: Molecular Luminescence Spectroscopy, *Pure Appl. Chem.* **56** (1984) 231–245.

20 Recommended Standards for Reporting Photochemical Data, *Pure Appl. Chem.* **56** (1984) 939–944.

21 Nomenclature of Inorganic Chemistry, Butterworths, London 1971.

22 Nomenclature of Organic Chemistry, Pergamon, Oxford 1979.

23 An Annotated Bibliography on Accuracy in Measurement, *Pure Appl. Chem.* **55** (1983) 907–930.

24 Assignment and Presentation of Uncertainties of the Numerical Results of Thermodynamic Measurements, *Pure Appl. Chem.* **53** (1981) 1805–1825.

25 Glossary of Terms Used in Physical Organic Chemistry, *Pure Appl. Chem.* **55** (1983) 1281—1371.

26 The Absolute Electrode Potential: An Explanatory Note, *Pure Appl. Chem.* **58** (1986) 955–966.

27 Electrode Reaction Orders, Transfer Coefficients and Rate Constants: Amplification of Definitions and Recommendations for Publication of Parameters, *Pure Appl. Chem.* **52** (1980) 233–240.

28 Nomenclature for Transport Phenomena in Electrolytic Systems, *Pure Appl. Chem.* **53** (1981) 1827–1840.

29 Bard, A.J., Parsons, R. and Jordan, J. (eds) *Standard Potentials in Aqueous Solutions*, Marcel Dekker Inc., New York 1985.

30 Definition of pH Scales, Standard Reference Values, Measurement of pH and Related Terminology, *Pure Appl. Chem.* **57** (1985) 531–542.

30a Reporting Physisorption Data for Gas/Solid Systems, *Pure Appl. Chem.* **57** (1985) 603–619.

Reporting Experimental Pressure-Area Data with Film Balances, *Pure Appl. Chem.* **57** (1985) 621–632.

Reporting Data on Adsorption from Solution at the Solid/Solution Interface, *Pure Appl. Chem.* **58** (1986) 967–984.

31 Atomic Weights of the Elements 1979, *Pure Appl. Chem.* **52** (1980) 2349–2384.

32 Element by Element Review of Their Atomic Weights, *Pure Appl. Chem.* **56** (1984) 695–768.

33 Atomic Weights of the Elements 1985, *Pure Appl. Chem.* **58** (1986) 1677–1692.

34 Isotopic Compositions of the Elements 1983, *Pure Appl. Chem.* **56** (1984) 675–694.

35 *Nomenclature of Inorganic Chemistry*. Chapters 1–3: Elements, Atoms, and Groups of Atoms (in preparation).

8.3 ADDITIONAL REFERENCES

36 Jenkins, F.A. Notation for the Spectra of Diatomic Molecules, *J. Opt. Soc. Amer.* **43** (1953) 425–426.

37 Mulliken, R.S. Report on Notation for the Spectra of Polyatomic Molecules, *J. Chem. Phys.* **23** (1955) 1997–2011. (Erratum *J. Chem. Phys.* **24** (1956) 1118.)

38 Herzberg, G. *Molecular Spectra and Molecular Structure.*
I. Spectra of Diatomic Molecules, Van Nostrand, Princeton 1950.
II. Infrared and Raman Spectra of Polyatomic Molecules, Van Nostrand, Princeton 1946.
III. Electronic Spectra and Electronic Structure of Polyatomic Molecules, Van Nostrand, Princeton 1966.

39 Watson, J.K.G. Aspects of Quartic and Sextic Centrifugal Effects on Rotational Energy Levels. In: Durig, J.R. (ed), *Vibrational Spectra and Structure*, Vol. 6, Elsevier, Amsterdam 1977, pp.1–89.

40 Callomon, J.H., Hirota, E., Kuchitsu, K., Lafferty, W.J., Maki, A.G. and Pote, C.S. Structure Data of Free Polyatomic Molecules. In: Hellwege, K.-H. and Hellwege A.M., (eds), *Landolt-Börnstein*, New Series, II/7, Springer-Verlag, Berlin 1976.

41 Bunker, P.R. *Molecular Symmetry and Spectroscopy*, Academic Press, New York 1979.

42 Brown, J.M., Hougen, J.T., Huber, K.-P., Johns, J.W.C., Kopp, I., Lefebvre-Brion, H., Merer, A.J., Ramsay, D.A., Rostas, J. and Zare, R.N., The Labeling of Parity Doublet Levels in Linear Molecules, *J. Mol. Spectrosc.* **55** (1975) 500–503.

43 Brand, J.C.D., Callomon, J.H., Innes, K.K., Jortner, J., Leach, S., Levy, D.H., Merer, A.J., Mills, I.M., Moore, C.B., Parmenter, C.S., Ramsay, D.A., Narahari Rao, K., Schlag, E.W., Watson, J.K.G. and Zare, R.N. The Vibrational Numbering of Bands in the Spectra of Polyatomic Molecules, *J. Mol. Spectrosc.* **99** (1983) 482–483.

44 Hahn, Th. (ed) *International Tables for Crystallography.* Vol. A, 2nd ed.: *Space-Group Symmetry*, Reidel Publishing Co., Dordrecht 1987.

45 Domalski, E.S. Selected Values of Heats of Combustion and Heats of Formation of Organic Compounds, *J. Phys. Chem. Ref. Data* **1** (1972) 221–277.

46 Freeman, R.D. Conversion of Standard (1 atm) Thermodynamic Data to the New Standard State Pressure, 1 bar (10^5 Pa):
Bull. Chem. Thermodyn. **25** (1982) 523–530;
J. Chem. Eng. Data **29** (1984) 105–111;
J. Chem. Educ. **62** (1985) 681–686.

47 Wagman, D.D., Evans, W.H., Parker, V.B., Schumm, R.H., Halow, I., Bailey, S.M., Churney, K.L. and Nuttall, R.L. The NBS Tables of Chemical Thermodynamic Properties, *J. Phys. Chem. Ref. Data* **11** Suppl. 2 (1982) 1–392.

48 Chase, M.W., Davies, C.A., Downey, J.R., Frurip, D.J., McDonald, R.A. and Syverud, A.N. JANAF Thermochemical Tables, 3rd ed. *J. Phys. Chem. Ref. Data* **14** Suppl. 1 (1985).

49 Glushko, V.P. (ed) *Termodinamicheskie svoistva individualnykh veshchestv.* Vol. 1–4, Nauka, Moscow 1978–85.

50 CODATA Task Group on Data for Chemical Kinetics, The Presentation of Chemical Kinetics Data in the Primary Literature, *CODATA Bull.* **13** (1974) 1–7.

51 Cohen, E.R. and Taylor, B.N. The 1986 Adjustment of the Fundamental Physical Constants, *CODATA Bull.* **63** (1986) 1–49.

52 Particle Data Group, Review of Particle Properties, *Phys. Lett.* **170B** (1986) 1–350.

53 Wapstra, A.H. and Audi, G. The 1983 Atomic Mass Evaluation. I. Atomic Mass Table, *Nucl. Phys.* **A432** (1985) 1–54.

54 Lederer, M. and Shirley, V.S. *Table of Isotopes*, 7th ed., Wiley Interscience, New York 1978.

THE GREEK ALPHABET

Α, α	A, α	*Alpha*
Β, β	B, β	*Beta*
Γ, γ	Γ, γ	*Gamma*
Δ, δ	Δ, δ	*Delta*
Ε, ε	E, ε	*Epsilon*
Ζ, ζ	Z, ζ	*Zeta*
Η, η	H, η	*Eta*
Θ, ϑ, θ	$\Theta, \vartheta, \theta$	*Theta*
Ι, ι	I, ι	*Iota*
Κ, ϰ	K, κ	*Kappa*
Λ, λ	Λ, λ	*Lambda*
Μ, μ	M, μ	*Mu*
Ν, ν	N, ν	*Nu*
Ξ, ξ	Ξ, ξ	*Xi*
Ο, ο	O, o	*Omicron*
Π, π	Π, π	*Pi*
Ρ, ρ	P, ρ	*Rho*
Σ, σ	Σ, σ	*Sigma*
Τ, τ	T, τ	*Tau*
ϒ, υ	Υ, υ	*Upsilon*
Φ, φ, ϕ	Φ, φ, ϕ	*Phi*
Χ, χ	X, χ	*Chi*
Ψ, ψ	Ψ, ψ	*Psi*
Ω, ω	Ω, ω	*Omega*

INDEX OF SYMBOLS

Qualifying subscripts, etc., are generally omitted from this index, so that for example E_p for potential energy and E_{ea} for electron affinity are both indexed simply under E for energy. The Latin alphabet is indexed ahead of the Greek alphabet. Symbols for units are included along with the symbols for physical quantities in alphabetical order.

Symbol	*Name*	*Page*
a	absorption coefficient	30
a	acceleration	13
a	activity	45, 51
a	are, unit of area	104
a	area per molecule	56
a_0	Bohr radius	19, 70, 81, 104
$\boldsymbol{a}, \boldsymbol{b}, \boldsymbol{c}$	lattice vectors	32
$\boldsymbol{a}^*$, $\boldsymbol{b}^*$, $\boldsymbol{c}^*$	reciprocal lattice vectors	32
a	specific surface area	56
a	thermal diffusivity	58
a, A	hyperfine coupling constant	24
amagat	amagat unit	107
atm	atmosphere, unit of pressure	48, 81, 106
A	absorbance	30
A	activity (radioactive)	20
A	ampere	65, 107
Å	ångström	23, 69, 104
A	area	13
A, $\mathscr{A}$	affinity of reaction	44
A, B	Einstein transition probabilities	29
A_H	Hall coefficient	33
A	Helmholtz energy	43
$\boldsymbol{A}$	magnetic vector potential	17
A	nucleon number, mass number	19
A	pre-exponential factor	49
A, B, C	rotational constants	22
A	spin–orbit coupling constant	22
A	van der Waals–Hamaker constant	56
Al	Alfvén number	59
AU	astronomical unit, unit of length	104
bar	bar, unit of pressure	48, 69, 106
b	barn, unit of area	69, 104
b	bohr	104
b	breadth	13
$\boldsymbol{b}$	Burgers vector	32
b	molality	38
b	mobility ratio	33
Bi	biot, unit of electric current	107
B	Debye–Waller factor	32
B	magnetic flux density	16
B	napierian absorbance	30
B	retarded van der Waals constant	56
B	rotational constant	20
B	second virial coefficient	44
B	susceptance	17
Btu	British thermal unit	106

Symbol	*Name*	*Page*
Bq	becquerel	66, 107
cal	calorie, unit of energy	106
c	amount concentration	38
c	speed, velocity	13, 29, 35
c_0	speed of light in vacuum	29, 81
cd	candela	65
c_1, c_2	first and second radiation constants	30, 81
C	capacitance	16
C	coulomb	66, 107
°C	degree Celsius	43, 66, 106
C	heat capacity	43
C	number concentration	35, 38
C_n	n-fold rotation operator	26
Ci	curie	107
Cl	clausius	106
Co	Cowling number	59
d	collision diameter	49
d	day	69, 105
d	degeneracy	23, 35
d	deuteron	38, 85
d	relative density	14
d	thickness, distance, diameter	13, 32
dyn	dyne, unit of force	105
D, d	centrifugal distortion constants	22
D	debye, unit of electric dipole	23, 108
D	Debye–Waller factor	32
D	diffusion coefficient	33, 58
D_{AB}	direct dipolar coupling constant	24
D	dissociation energy	19
$\boldsymbol{D}$	electric displacement	16
Da	dalton (unified atomic mass unit)	19, 37, 69, 105
e	base of natural logarithms	75, 82
e	electron	38, 85
e	elementary charge	19, 51, 70, 81, 107
e	linear strain	14
erg	erg, unit of energy	105
e.u.	entropy unit	107
eV	electronvolt, unit of energy	69, 106
$\boldsymbol{E}$	electric field strength	16
E	electromotive force	16, 51
E	energy	14, 19, 33, 49
E_{h}	hartree energy	70, 81, 105, 114
E	identity symmetry operator	26
E	irradiance	30
E	modulus of elasticity	14
E	thermoelectric force	33
Eu	Euler number	58
f	activity coefficient	45
f	fermi, unit of length	104
f	frequency	13
f	friction coefficient	15
f	fugacity	45
$f(c)$	velocity distribution function	35
f, F	vibrational force constant	23
ft	foot	104

Symbol	*Name*	*Page*
F	farad	66
F	Faraday constant	51, 81
$\boldsymbol{F}$	force	14
F	rotational term (spectroscopy)	22
$F(c)$	speed distribution function	35
F	total angular momentum	25
Fo	Fourier number	58
Fr	franklin, unit of electric charge	107, 112
Fr	Froude number	58
g, g_n	acceleration (due to gravity)	13, 81, 105
g	degeneracy	23, 35
g	density of vibrational modes	32
g, g_e	Landé g-factor	20, 24, 81
g	vibrational anharmonicity constant	22
g	gram	68, 105
gal	gallon	105
gr	grain, unit of mass	105
grad	grade	107
G	electrical conductance	17
G	gauss	109
G	Gibbs energy	43
G	gravitational constant	14, 81
$\boldsymbol{G}$	reciprocal lattice vector	32
G	shear modulus	14
G	thermal conductance	58
G	vibrational term (spectroscopy)	22
G	weight	14
Gal	gal, galileo	105
Gr	Grashof number	58
Gy	gray	66, 107
h	coefficient of heat transfer	58
h	film thickness	56
h	height	13
h	helion	38, 85
h	hour	69, 105
ha	hectare, unit of area	104
hp	horse power	106
h, k, l	Miller indices	34
$h, \hbar$	Planck constant ($\hbar = h/2\pi$)	19, 29, 70, 81
H	enthalpy	43
H	hamilton function	14, 18
H	henry	66
$\boldsymbol{H}$	magnetic field	16
Ha	Hartmann number	59
Hz	hertz	13, 66
i	square root of minus one	76
in	inch	104
i	inversion operator	26
I	electric current	4, 16, 52
I	ionic strength	45, 51
I	moment of inertia	14, 22
I	nuclear spin angular momentum	25
I	radiant intensity (luminous intensity)	4, 29
$\boldsymbol{j}, \boldsymbol{J}$	electric current density	16, 18, 52
j, J	angular momentum	25

Symbol	*Name*	*Page*
J	flux	58
J_{AB}	indirect spin–spin coupling constant	24
J	joule	66, 105
J	Massieu function	43
J	moment of inertia	14
k	absorption index	30
k	Boltzman constant	35, 49, 81
k_d	mass transfer coefficient	58
k	rate constant	49, 52
k	thermal conductivity	58
kg	kilogram	65, 105
k_{rst}	vibrational force constant	23
$\mathbf{k}$	wave vector	33
kgf	kilogram-force	105
k, K	component of angular momentum	25
K	absorption coefficient	30
K	bulk modulus	14
K_{cell}	conductivity cell constant	53
K	equilibrium constant	44
K	kelvin	65, 106
K	kinetic energy	14
Kn	Knudsen number	58
l, L	angular momentum	14, 23, 25
l, L	length	4, 13, 33
l, L	litre	69, 105
lb	pound	105
lm	lumen	66
lx	lux	66
l.y.	light year	104
L	Avogadro constant	35, 37, 81
L	inductance	17
L	Lagrange function	14
Le	Lewis number	59
$\mathbf{m}$	magnetic dipole moment	17, 19
m	mass	4, 14, 19, 33, 37, 70
m_e	electron rest mass	19, 70, 81
m_n	neutron rest mass	81
m_p	proton rest mass	81
m_u	atomic mass constant	19, 81
m	metre	65, 104
m	molality	38, 51
m, M	component of angular momentum	25
mi	mile	104
min	minute	69, 105
mol	mole	4, 41, 65
mmHg	mm of mercury, unit of pressure	106
$\mathcal{M}$	Madelung constant	32
$\mathbf{M}$	magnetization	16
M	molar, unit of concentration	38
M	molar mass	37, 56
M	mutual inductance	17
M	radiant exitance	29
M	transition dipole moment	23
M	torque	14
Ma	Mach number	58
Mx	maxwell, unit of magnetic flux	109

Symbol	*Name*	*Page*
R	gas constant	35, 81
R_{H}	Hall coefficient	33
$\boldsymbol{R}$	lattice vector	32
R	molar refraction	30
R	nuclear orbital angular momentum	25
$\boldsymbol{R}$	position vector	32
R	röntgen	107
R, R_{∞}	Rydberg constant	19, 81
Re	Reynolds number	58, 59
Rm	magnetic Reynolds number	59
Ry	rydberg, unit of energy	105
R	thermal resistance	58
$\boldsymbol{R}$	transition dipole moment	23
s	length of path, length of arc	13
s	long-range order parameter	32
s	second	65, 105
s	sedimentation coefficient	56
s	specific surface area	56
s	solubility	38
s	symmetry number	36
sr	steradian	13, 66
s, S	spin angular momentum	25
S	area	13
S	entropy	43
S_{AB}	overlap integral	18
S	probability current density	18
$\boldsymbol{S}$	Poynting vector	17
S_n	rotation–reflection operator	26
S	siemens	66
S	symmetry coordinate	23
Sc	Schmidt number	59
Sh	Sherwood number	59
Sr	Strouhal number	58
St	Stanton number	58
St	stokes	106
Sv	sievert	66, 107
Sv	svedberg, unit of time	105
t	film thickness, thickness of layer	56
t	Celsius temperature	43
t	time	4, 13, 21, 49
t	tonne	69, 105
t	transport number	53
t	triton	38, 85
T	tesla	66, 109
$\boldsymbol{T}$	hyperfine coupling tensor	24
T	kinetic energy	14, 18
T	period, characteristic time interval	13, 20, 21, 24
T	term value (spectroscopy)	22
T	(thermodynamic) temperature	4, 33, 43, 55
T	transmittance	30
T	torque	14
Torr	torr, unit of pressure	106
u	Bloch function	33
u	speed, velocity	13, 35
u	electric mobility	53

Symbol	*Name*	*Page*
u	unified atomic mass unit	69, 81, 105
U	electric potential difference	16
U	internal energy	43
v	rate of reaction	49
v	specific volume	14
v	velocity	13
v	vibrational quantum number	22
V	electric potential	16, 51
V	potential energy	14
V	volt	66, 108
V, v	volume	13, 14, 37
w	energy density	29
w	mass fraction	37
w	speed, velocity	13
w, W	work	14, 43
W	watt	66, 106
W	radiant energy	29
W	statistical weight	35
W	weight	14
Wb	weber	66, 109
We	Weber number	58
x, y, z	cartesian coordinates	13
x, y	mole fraction	37
x	vibrational anharmonicity constant	22
X	reactance	17
X	x unit	104
yd	yard	104
Y	admittance	17
Y	Planck function	43
z	charge number of an ion	51
z	collision frequency factor	50
z, Z	partition function	35
Z	compression factor	44
Z	collision frequency, collision number	49
Z	impedance	17
Z	proton number, atomic number	19
α	absorptance	30
α	absorption coefficient, napierian	30
α	acoustic absorption factor	15
α	alpha particle	38, 85
α	angle of optical rotation	30
α	degree of dissociation	38
α	electric polarizability	20
α	electrochemical transfer coefficient	52
α	expansion coefficient	43
α	fine structure constant	19, 70, 81
α	Madelung constant	32
α	transfer coefficient	52
$\alpha, \beta, \gamma, \ldots$	plane angle	13
α_p	relative pressure coefficient	43
α, β	spin wavefunctions	18
β	pressure coefficient	43
β	reciprocal temperature parameter, $1/kT$	36
β	retarded van der Waals constant	56

Symbol	*Name*	*Page*
μ	chemical potential	44, 52
$\boldsymbol{\mu}$	electric dipole moment	16, 20, 23
μ	electric mobility	53
μ	friction coefficient	15
μ	Joule–Thomson coefficient	43
$\boldsymbol{\mu}$	magnetic dipole moment	17, 19
μ_B	Bohr magneton	20, 70, 81, 109
μ_e	electron magnetic moment	19, 81
μ_N	nuclear magneton	20, 81, 109
μ_p	proton magnetic moment	19, 81
$\mathrm{\mu}$	micron	104
$\mathrm{\mu}$	muon	38, 85
μ	mobility	33
μ	permeability	16
μ_0	permeability of vacuum	16, 81, 111
μ	reduced mass	14
μ	Thomson coefficient	33
μ	viscosity	15
ν	frequency	13, 20, 22, 29
ν	kinematic viscosity	15
$\mathrm{\nu}_\mathrm{e}$	neutrino	85
ν	stoichiometric number	38
$\tilde{\nu}$	wavenumber in vacuum	29
ξ	extent of reaction, advancement	38, 49
ξ	magnetizability	19
Ξ	grand partition function	35
$\boldsymbol{\pi}$	angular momentum	25
$\mathrm{\pi}$	circumference/diameter	82
$\mathrm{\pi}$	pion	85
π	surface pressure	56
Π	osmotic pressure	45
Π	Peltier coefficient	33
$\mathrm{\Pi}$	product sign	75
ρ	mass density, mass concentration	14, 37
ρ	density of states	35
ρ	acoustic reflection factor	15
ρ	charge density	16, 18, 33
ρ	energy density	29
ρ	reflectance	30
ρ	resistivity	17, 33
ρ_A, ρ_S	surface density	14
σ	area per molecule	56
σ	electrical conductivity	17, 33, 52
σ	cross section	21, 49
σ	normal stress	14
σ	reflection plane	26
σ	shielding constant (NMR)	23
σ	short-range order parameter	32
σ	Stefan–Boltzmann constant	30, 81
σ	surface tension	14, 43, 56
σ	surface charge density	16, 52
σ	symmetry number	36
σ	wavenumber	29
σ, Σ	axial spin angular momentum	25
Σ_f	film tension	56
$\mathrm{\Sigma}$	summation sign	75

PRESSURE CONVERSION FACTORS

		Pa	*kPa*	*bar*	*atm*	*mbar*	*Torr*
1 Pa	=	1	10^{-3}	10^{-5}	$9.869\,23 \times 10^{-6}$	10^{-2}	$7.500\,62 \times 10^{-3}$
1 kPa	=	10^3	1	10^{-2}	$9.869\,23 \times 10^{-3}$	10	7.500 62
1 bar	=	10^5	10^2	1	0.986 923	10^3	750.062
1 atm	=	101 325	101.325	1.013 25	1	1013.25	760
1 mbar	=	100	10^{-1}	10^{-3}	$9.869\,23 \times 10^{-4}$	1	0.750 06
1 Torr	=	133.322	0.133 322	$1.333\,22 \times 10^{-3}$	$1.315\,79 \times 10^{-3}$	1.333 22	1

Examples of the use of this table:

1 bar = 0.986 923 atm
1 Torr = 133.322 Pa

Note: 1 mmHg = 1 Torr, to better than 2×10^{-7} Torr (see p.106).